Fields That

Fields That

Dream

A Journey to the Roots of Our Food

Jenny Kurzweil

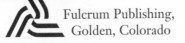
Fulcrum Publishing,
Golden, Colorado

Since our break with nature came with agriculture, it seems fitting that the healing of culture begin with agriculture, fitting that agriculture take the lead.
—*Wes Jackson*, Becoming Native to This Place

Library of Congress Cataloging-in-Publication Data
Kurzweil, Jenny.
    Fields that dream : a journey to the roots of our food / by Jenny Kurzweil.
        p. cm.
    Includes bibliographical references.
    ISBN 1-55591-506-X (pbk.) 978-1-55591-506-3 1. Farm life—Washington
(State)—Anecdotes. 2. Agriculture—Washington (State)—
Anecdotes. 3. Farmers—Washington (State)—Anecdotes. I. Title.
    S521.5.W2K87 2005
    630'.9797—dc22

                            2005015983

ISBN10 1-55591-506-X ISBN13 978-1-55591-506-3
Printed in the United States of America
0 9 8 7 6 5 4 3 2 1

Editorial: Faith Marcovecchio, Haley Groce
Design and cover image: Jack Lenzo

Wes Jackson quote from *Becoming Native to This Place*, reprinted by permission of University Press of Kentucky, Lexington, Kentucky.

Gloria Anzaldúa quote from *Borderlands: La Frontera: The New Mestiza*,
reprinted by permission of Aunt Lute Books, San Francisco, California.

Nancy D. Donnelly quote from *Changing Lives of Refugee Hmong Women*,
reprinted by permission of the University of Washington Press.

Fulcrum Publishing
16100 Table Mountain Parkway, Suite 300
Golden, Colorado 80403
(800) 992-2908 • (303) 277-1623
www.fulcrum-books.com

For Andrea and for Jacob
—a brand new day

100 percent of the royalties of *Fields That Dream: A Journey to the Roots of Our Food* will be split between the following organizations:

The Neighborhood Farmers' Market Alliance

The Neighborhood Farmers' Market Alliance (NFMA) is a community-based organization developed in response to the growing popularity and public support of the neighborhood farmers' markets in Seattle. The NFMA's mission is to support Washington's small farms and farming families by providing effective direct sales sites for the region's small farmers and by educating consumers about farm products and the benefits of buying direct from local farmers.

Neighborhood Farmers' Market Alliance
4519 ½ University Way NE
Ste. 202
Seattle, WA 98105
(206) 547-2278
nfma@seattlefarmersmarkets.org

PCC Farmland Fund

The Farmland Fund works to secure and preserve threatened farmland in Washington State and move it into organic production. The fund's focus is on large, functional landscapes of local, regional, and statewide importance so protection can be extended to biodiversity and wildlife habitat as well as to farmers and farming communities. The Farmland Fund is an independent, community-supported 501(c)3 land trust. It was founded in 1999 by PCC Natural Markets as a separate, nonprofit organization.

PCC Farmland Fund
4201 Roosevelt Way NE
Seattle, WA 98105
(206) 547-1222, ext. 140
farmlandfund@pccsea.com

Table of

Contents

Acknowledgments

People say that writing a book is like having a baby. It isn't. But it is like raising a child—you know, that cliché saying "It takes a village to raise a child." Except in my case, writing this book has taken more like a small metropolis.

Fields That Dream could not have been written without the generosity of all those I interviewed—Chris Curtis and the farmers at the University District Farmers' Market in Seattle: Margaret Hauptman, Gretchen Hoyt, John Huschle, Juana Lopez, Joanie McIntyre and Michael Shriver, the late Robert Meyer, Jeff Miller, Steve and Beverly Phillips, Andrew Stout, Katsumi and Ryoko Taki, Joua Pao Yang, Xee Yang-Schell, and Susan Wells. Thank you for opening your homes, sharing your stories, and taking the time to read and reread what I had written about you. Thank you to other Washington State farmers who shared their insight and time: Hilario Alvarez, Linda Bartlett, Fong Cha, Kayeng Cha, Martha Goodlett, Gerald and Anne Goronea, Steve Hallstrom, and Maika Xiong.

Thank you to those who shared their expertise: Mark Musick for his extensive knowledge about sustainable agriculture in the Pacific Northwest; Dr. Richard Kirkendall for his help on the history of agriculture; Dr. Kim Waddell for educating me on the science of genetically modified organisms; Marilú Chávez for her flawless translation of Juana Lopez's story into Spanish; and the librarians at the Ballard Public Library, who took the time to help a research rookie and introduced me to the joys of Google.

Anna Davidson, thank you for taking such exquisite photographs that capture the beauty and commitment of the farmers at the University District Farmers' Market.

Thank you to Hedgebrook and my Hedgebrook sisters of January 2001 for making me feel like a real writer and providing such a sacred and concentrated space to write.

Thank you to Venerable Thubten Chodron and Venerable Robina Courtin for teaching the importance of setting motivation and introducing the idea that writing can be dharma practice.

Thank you to my writing friends Gretchen Primack, Jennifer Munro, and Ayize Jama-Everett, who provided much-needed encouragement and advice throughout the long journey of writing, editing, and submitting this manuscript. And sincere gratitude for the support of my entire extended family of friends.

Thanks to Marlene Blessing, former editor at Fulcrum Publishing, for her warmth, enthusiasm, and belief in the project. Much gratitude to Sam Scinta, vice president, and Faith Marcovecchio, editor, at Fulcrum who saw the project through to the end with courtesy, patience, and efficiency, and both of whom pushed me (gently) to do much-needed edits beyond what I thought possible.

Thank you to Dr. Sue-Ellen Jacobs, my mentor and friend, who guided me with a firm and loving hand into the world of research, writing, and oral history.

Deep thanks to my father, Jack Kurzweil, and late grandfather Herbert Aptheker, who generously provided me with much-needed financial assistance that allowed me to complete the first draft of the book.

A universe of gratitude to my mothers, Bettina Aptheker and Kate Miller—without their love and guidance, neither this book, nor I, would be here.

Fields That Dream

And to Andrea Roth, who has lived and breathed this project with me for the last six years, thank you for holding fast, for holding me up, and for believing that it was possible when I never did.

Acknowledgments

Introduction:

Beginnings

As a kid, I knew that almost everything I ate was grown just minutes away from my home, and I think I assumed it was that way for everyone. Pacific Grove, where I grew up, is a little town on the central coast of California, just a breath away from the giant Salinas Valley, one of the largest areas of agricultural production in the country. There are vast farms filled with lettuce, artichokes, garlic, carrots, strawberries, and Brussels sprouts. I saw the fields pass by endlessly, with rows and rows of people bent over harvesting, whenever we took the highway out of town. I knew that the farmworkers were mostly Mexicans and that they lived in Salinas and other neighboring towns.

Pacific Grove was mostly white and upper middle class, and the rest of the peninsula largely segregated, including the almost completely Mexican-American city of Salinas. I remember once in high school when my class was on a bus driving past a strawberry field. A teacher looked out the window, pointing to the lines of people bending down picking fruit, and laughed, "You all better study hard, because if you don't, you will end up out there!"

Summer camp was what I looked forward to most every year. It was a working ranch and organic farm on the floor of a little valley, including more than a thousand acres of wilderness preserve in the surrounding Los Altos Hills. Hidden Villa was everything that Pacific Grove wasn't: there were campers from a wide range of racial and financial backgrounds and a very diverse staff. I first went to camp when I was ten, and at fifteen I started helping in the camp kitchen. Over the following four summers, cooking developed into my main passion. By the time I was eighteen, I was co–head cook, buying, planning, and cooking three meals a day for 120 people. I loved the responsibility, independence, and physical hard work of running a kitchen.

In the kitchen we used to plan our menus after consulting the farmer about which produce would be available when. It just made sense to use the vegetables that were being grown outside the back door, and it was fun to put enormous pots of water on the stove to boil and then send someone out to pick the corn. The farm didn't produce everything we needed, so we ordered the rest. The boxes that landed on our loading dock were all from farms within a 100-mile radius.

I didn't give any of this another thought until the next summer, when, at nineteen, I went to Vermont to live on a friend's organic farm and work as a baker and line cook at a local restaurant. I was shocked on my first day when I checked in the produce delivery. The carrots were from Salinas, the strawberries from Watsonville, the garlic from Gilroy, and so on. I couldn't believe it. My friend was growing all of these things right down the road, and the restaurant was getting them from clear across the country—from the fields just outside the town where I grew up. One of the chefs at the restaurant had worked in a resort in the Caribbean, and he said that even there the strawberries they used were from Watsonville. Maybe he was pulling my leg, but it seems absurd enough to be true.

That summer in Vermont was the first time I was truly exposed to life outside my 120-mile coastal California cocoon. The revelation that food was being shipped across the country

Fields That Dream

when it was available just around the corner made me realize that while I was so focused on cooking, I had never thought about what exactly I was feeding myself, where it was really coming from, and who was actually growing it. Looking back, I realized that none of the fields near my home had signs that announced their names. There were no farm tours or invitations to harvest parties. There were just miles and miles of anonymous vegetables.

My experience in Vermont began a more than ten-year journey to learn about the roots of my food. From Vermont I moved to Seattle to attend the University of Washington. To support myself through school, I continued cooking, but became anxious to learn the origins of the food I was purchasing, cooking, and selling to people. What I have learned is shocking. From pesticide use to development of farmland to the corporate takeover of family farms, there is so much more behind the produce at the grocery store than I ever imagined.

Perhaps most terrifying is the seemingly out-of-control use of pesticides. There are 800 chemicals approved for use as pesticides in the United States, and from this core group, more than 20,000 pesticide products are marketed with more than 4.5 billion pounds used annually in our country.[1] Despite these huge amounts, the percentage of crops lost to pests has increased nearly 20 percent since the end of World War II.[2] The Environmental Protection Agency (EPA) has classified 250 of those 800 chemicals as carcinogenic (cancer causing), and twelve of those chemicals are some of the most widely used pesticides in the country—over 380 million pounds of these known cancer-causing chemicals are applied to U.S. fields every year.[3] The EPA estimates that tens of thousands of farmworkers suffer from pesticide exposure each year, and pesticides are

Beginnings

linked to infertility, childhood leukemia, and brain cancer. The Pesticide Reduction Initiative reports that "Pound for pound, children not only have higher exposure to pesticides than adults do, but also are more vulnerable—their developing systems are more susceptible to the neurotoxic effects of pesticides."[4] In their 1993 report on pesticides in the diets of infants and children, the National Research Council explained that more than 143,000 American two-year-olds consume ten times the EPA's acceptable level of organophosphates, a class of pesticides.[5]

And pesticides are just the beginning of the story. More than 1 million acres of American farmland are developed each year, and that rate is increasing.[6] The development of farmland is a concern on many levels. Not only does the food and farming system employ nearly 23 million Americans, farmland creates the economic core of thousands of rural and urban communities across the country.[7] Farmland is also essential in the environmental sustainability of the country. Aside from the nice scenery, if managed properly, farmland can provide stable ecosystems for wildlife, watershed management, and flood control.[8]

This rate of farmland development, along with the fact that it is increasingly difficult to make a living as a farmer, means that the number of farms has plummeted. According to the United States Department of Agriculture (USDA), in 1936 there were 6.8 million farms. In 1997 there were only 1.9 million, and 94 percent of these existing farms made less than $250,000 per year. The overwhelming majority of these farms (86 percent) are owned by families or individuals. This means that the remaining 6 percent of the farms are making all of the money, and they are owned by corporations, institutions, and partnerships.

These current agricultural chemical uses and farmland development practices are simply not sustainable. How much

Fields That Dream

farmland will we have left in ten years? On the farmland that does exist, how toxic will the soil be? How are individual- and family-owned farms going to survive in the face of corporate takeover? What will it take to create a sustainable future?

The process of learning where my food comes from was leaving me with more questions than when I started. It was like opening a huge can of worms—pesticides linked to policies and policies linked to lobbyists and lobbyists linked to companies that are owned by other companies that are owned by corporations. At the same time that I knew I needed to learn more, I was afraid to do so, frightened by my sense of powerlessness in the face of the intricate network of commerce and legislation.

One Saturday morning, a friend took me to the University District Farmers' Market in Seattle, and amidst the bounty of fresh vegetables, fruit, and flowers, my fear began to lessen. That day at the market I saw evidence that there is indeed a better way to grow our food. I quickly became a U-District Farmers' Market regular.

I saw farmers show up every Saturday with truckloads of produce grown within a 100-mile radius of the city with little or no pesticides and people braving enormous amounts of rain to buy those vegetables, fruits, and flowers. It all made me feel a little more at home with the world. I began to believe that I had some breathing room on a planet where we are constantly bombarded with bad news.

Aptly named, the market is right in the middle of Seattle's University District, which has its share of student hangouts, big rambling houses, and a treasure trove of used bookstores. "The Ave," University Street's local nickname, is the heart and soul of the district and frequented by students, professors, and everyone in between. There are venerable churches, crowded coffeehouses, and the University Heights Community Center, which offers a wide range of activities such as an art school and day care center. In the shadow of this very old, large tan building, behind a chain-link fence strewn with paintings of brightly colored vegetables and butcher paper signs, farmers, merchants, and thousands of shoppers converge each Saturday morning five months out of the year.

The market is truly a destination. Live music fills the air, local chefs demonstrate cooking techniques, and a teenage Shakespeare company puts on abbreviated plays in the parking lot next door. People who don't even know each other talk and laugh together. I can overhear snippets of conversations about the lovely goat cheese from Port Madison Farm and how marvelous Willie Greens' arugula is. Strangers tell me about the delicious pie they made with last week's blueberries and ask if I have any idea what to do with kabocha squash. While the market is loud and a bit chaotic, there is an intimacy to these interactions that is unusual in a city of well over half a million people.

The University District Farmers' Market is a prime example of what is happening across the country. The USDA reports that there are more than 3,100 farmers' markets throughout the United States, 79 percent more than there were in 1994. Many of the farmers selling at these markets grow organically, and they are benefiting from America's conscious choice to start buying organic. (Organic food is produced without the use of pesticides, synthetic fertilizers, biotechnology, and ionizing radiation.) In fact, the Organic Trade Association reports that sales of organic food in the United States grew from $1 billion in 1990 to $7.7 billion in 2001.

When I buy food at the farmers' market, I know that it has not been shipped back and forth across the country. It has not

Fields That Dream

been grown by multinational corporations, but by families. Most importantly, when I buy food at the farmers' market, I meet the grower. I have a connection, an interaction, and a place to express my gratitude.

I have wanted to know how to make sense out of a world that teaches our children to mock farmworkers as they toil to bring food to our tables. I have wanted to know the true origins of my food. Through meeting the farmer I have gained the courage to learn more. I know their names now: Bob, the retired organizer, who is passionate about land preservation; Susan, a former insurance broker turned sprout farmer; and Andrew, the idealistic entrepreneur. There is Joanie, a hardworking mother of four and Xee, who bridges generations, learning to farm in the traditional ways of her family.

Here are the stories of the people who feed the city of Seattle. They are a wildly diverse group of farmers growing food in a better way, be it for preservation of the earth, health of its inhabitants—both human and non—or a better life for themselves, their communities, and communities at large. Together they create a sustainable future for the Pacific Northwest.

But the story of the family farmer is not just a local or regional one. It extends to the tanned, work-lined faces of the farmers at any market in the country. These people *choose* to work in small-scale farming, to be stewards of the land. In these pages, they tell what drives them to eke out a living from the soil, what they fight against, and what they believe in.

Their lives show us how we moved from a country of small farms to a landscape scarred by agricultural corporations. Their struggle informs us about biotechnology and urban sprawl, immigration laws, and international trade. And their stories give us hope for the future.

Beginnings

Chapter One: Pioneer

I'm driving out to Tenino, Washington, to visit Bob and Pat Meyer at Stoney Plains Farm. The town of Tenino is twenty minutes south of Olympia, the state capital, and about an hour and a half from Seattle. Bob told me on the phone that he is actually a ways outside of town, but I've come a fair distance from the main highway and am beginning to get nervous that I am lost. This doesn't really look like farm country.

A strange congregation of houses lines the road. It seems like this area may have been farmland years ago, but from what I can see, there isn't enough space between the hodgepodge of houses for much more than a midsized garden. Although it is still too rural to be the suburbs, from the concentration of houses it appears that Tenino is on its way to becoming a bedroom community to Olympia. I pass conservative-looking ramblers with gnomes and pink flamingos that keep vigilant watch over passersby and derelict homes with rusted-out farm equipment and abandoned Datsuns sprouting out of the earth like weeds. There are a very large number of brand-spanking-new houses, white, bright, and big, complete with two-car garage and minivan parked in the driveway.

Finally I see the turn to the Meyers' place, and as I approach the farm, the empty space spills refreshingly out before me. From what I have seen, it looks like I am approaching one of the last small farms in the area.

The Meyers have always been farmers at heart. Even

when they lived in big city St. Paul they were farming, albeit on a small scale. As we sit at their kitchen table, Bob explains how he and his wife gardened up every inch of soil they could claim in St. Paul's community gardens.

Bob chuckles, "It was funny. We used to get two in my name, two in the wife's name, two in the neighbor's name ... We had plots all over town! We raised food for our family and friends. It was kind of a hobby. I often teased my seven kids that it was my way of taking my frustrations out—on the weeds rather than on them!"

The Meyers have come a long way from gardening plots scattered around a city to the twenty-five-acre Stoney Plains Farm. Their city gardening days began when Bob worked for a national labor union based out of Minnesota.

He explains further, " I did a lot of organizing and I got involved every time we had a big strike or shoot-out with an employer." Bob leans across the table and says, smiling mischievously, "I have been known to be a soapbox liberal type."

I could see how Bob would have made a good organizer by just his appearance alone. He looks like a force to be reckoned with, at an imposing six-foot-something. He is broad shouldered with a healthy belly and long legs. At sixty, Bob has a slightly graying mustache, a receding hairline, jowly cheeks, and a great booming laugh. A baseball cap sits lightly on his head.

When Bob was transferred from Minnesota to Washington, he and Pat bought a house on two lots in Lacey, just north of Olympia. They had room for a sixty-by-eighty foot garden that was planted intensively. In 1978 they started selling their extra produce on Saturdays at the Olympia farmers' market. Back then it was just a hobby. The Meyers bought the Tenino piece of property before they actually lived on it, farming it in addition

Fields That Dream

to their Lacey garden. They didn't have any equipment, and the first ten acres they did all of the tilling, planting, and harvesting by hand. Pat worked on the farm full time, and, after he retired, Bob started in full time as well.

He says earnestly, "I made up my mind a long time ago that if I ever retired, I wasn't going to sit in the rocking chair. Number one, it would drive me mentally nuts, and number two, it would probably kill me."

It took Bob a while to decide to retire fully—union work seems to be in his blood. "You see," he says, settling back in his chair, "my dad was a printer, and back in those days the printers were all unionized. So I guess the union work is just kind of inherited. I definitely enjoyed my work at the union. I never had a day that I didn't want to go to work." Then Bob lets out a sigh, shrugs his massive shoulders, and says, "I am sitting here wondering what is the matter with unions today. But I stay out of it. I think they're missing the boat, but what do I know? Anyway, I just liked organizing, it was a lot of fun. I had a lot of fun stirring up shit."

Now instead of stirring it up, Bob is laying shit out in his fields. As we are talking, a big truck pulls up and the driver starts maneuvering his way onto the narrow driveway. Bob explains that he is from the local chicken plant and has come to spread manure over the spring fields.

It is a gray day in late April, and there is still a very soggy chill to the air. Bob and I don our coats and head outside so he can give instructions to the truck driver and then show me around the property. I soon find out that Stoney Plains Farm is an impressive operation. Starting out with the initial ten acres, it didn't take long for Bob and Pat to accumulate another fifteen. They are now looking for an additional ten or so acres.

He laughs, "At the same time I am saying that I am going to cut back!"

Stoney Plains produces a serious amount of food. They sell produce from May until December. Eighty percent of their produce is sold at farmers' markets five days a week; the rest is sold wholesale. Bob nods his head, "That's our bread and butter, the farmers' markets. And they're fun too."

Bob tells me that at the height of the season he has about ten people working for him and that it has been essentially the same crew for seven or eight years. "They are all local Hispanic folks," he says, "though sometimes some of their relatives come up from Mexico. It is all very much a family affair."

We have reached the door to the first of three greenhouses, and Bob pauses to turn the knob and let me in. Greeted by a rush of warm air and a rich earthy smell, I am delighted by the wash of color all around me as summer explodes three months early. As I marvel at the expanse of geraniums and New Guinea impatiens, Bob continues his thought about the loyalty of his crew.

He says seriously, "You know, it's theirs. They feel that the farm is theirs. One of the rules that we have always had with our crew is when you work here, anything [is yours] for your table. Not for the neighbors or for canning, but anything for your table. You just take it. I don't know if I am a great person to work for. We sure as hell can't pay much money, but we treat them with dignity, and I think that means a lot."

Bob shows me around the greenhouse, and he seems to feel that since we are there, he should do a little work. He steers me over to a workbench and starts planting lettuce seed into starting trays that are filled to the brim with a dark, almost delicate-looking soil. The seeds he is planting look like the little pink candies you get after dinner in an Indian restaurant. Bob

Fields That Dream

explains that this light protective coating makes them easier to handle for mass planting. He tells me that in some of the seed produced, the protective coating actually includes a spot of pesticide or fungicide.

"Although this seed is organic," Bob says, his voice full of concentration as he sifts the tiny seeds through his broad fingers and into their delegated trays, "some of the seeds that we get are treated. We are a certified organic farm. But the rule says that you can only use treated seed if the nontreated is not available. Of course, the seed houses understand this; if the untreated seed is not available, all they have to do is send a letter. Big deal. So I order three thousand dollars worth of seed, and the seed comes, and I get a letter that says this, this, and this is treated and ... "

Bob looks up, seemingly frustrated by all of the bureaucratic red tape that even the organics industry is not immune from. "What does it solve? Nothing. I mean, I'd rather not use treated seed, but I'd really rather focus on cover crops, natural manures, you know, some of the stuff that is really important, as opposed to a microscopic speck of pesticide or what have you, that probably don't mean a hill of beans as far as affecting the environment."

Bob shuffles through some seed packages and then starts planting a different kind of lettuce, explaining the process to me as he does it. I offer to help, but Bob motions for me to relax, so I lean against the greenhouse wall and watch him work.

After a few minutes, his hands are empty. Bob lifts his baseball cap to scratch his head. "I think if you are going to be a sustainable farmer, number one, you have to be willing to experiment. Number two, be willing to gamble. Maybe that says the same thing, I don't know.

"The real difference between organic and conventional farming is dealing with the weeds. It is not even pests, it is weeds. I spend more money on getting rid of weeds than I do any other function on the farm. Now, if I was commercial, I could go spray them once, it would cost me forty-five dollars an acre, and I would have a clear field. I will put six people in that same field and they will work the better part of the week to get it clean. That is the difference. But all the work is worth it in the long run."

Bob gathers up all of the seed packs into a bag and then picks up each tray he has planted and lightly taps its bottom onto the work surface. The soil settles with a sigh and envelopes the pink seeds, which in two months will be the big heads of red lettuce that I buy from him at the market.

We are greeted with a gust of cold wind as we leave the greenhouse and head back toward the house. When we enter, we are met by a very cheerful-looking woman with curly gray hair, big rosy cheeks, and a pink, embroidered sweatshirt. Bob introduces me to his wife, Pat, who promptly takes me on a tour of the house. Eventually Pat and Bob lead me to the basement, where they have a two-bedroom apartment for one of their sons and space for Bob's shop and food storage. The shelves in the storage space are jam-packed with canning jars full of applesauce, peaches, tomatoes, beans, vegetable medleys, and countless other fruits and vegetables. Bob then points out two freezers which Pat tells me are full of melons, berries, squash, and pork from pigs they have raised themselves.

When we head back upstairs, I ask Pat if she was surprised to see how she went from a backyard garden in Lacey to a twenty-five-acre farm.

She answers in her efficient yet kind tone, "No. Bob's

Fields That Dream

mother always told me that he had wanted a farm ever since he was a little kid. It's just I never figured that I would be doing the farming while he was working!" She laughs as she says good-bye and heads out to visit friends.

Bob gets himself a cup of coffee and I shake my head when he offers me some. He settles down at the kitchen table across from me and takes a small slurp from the mug. There is a deck of cards on the table, and Bob shuffles them as he explains what they do with the produce they don't sell at the farmers' market.

"Of course, in Seattle there is the University District Food Bank that comes to the Saturday market. But down here in Olympia, we tried for years to get the food bank, or anyone for that matter, to come. But we couldn't get anybody with any consistency."

After much searching, Bob found a Baptist church that runs a soup kitchen. He nods in satisfaction, "That was eight years ago, and they haven't missed a pickup since. They are feeding right now, every Sunday, about 400 people from the street. Everybody from kids in diapers to seniors."

Bob laughs, "I have this one friend who is a fruit vendor at the Olympia market, and he gives the church pallets of apples. What they can't use they'll freeze. So once I ask him, 'Hey, if you've got some apples left over, the wife wants to make some applesauce.' He said, 'Yeah, well, I think I'll have some after I take care of the church!'"

As I listen to Bob, I look again at him and my surroundings. I realize that Bob and Pat do not really fit into my preconceived notion of organic farmers. I don't know why. I guess they are just so all-American. Even their kitchen has ruffled curtains and a cross-stitch sampler hanging on the wall. They have enough food downstairs to get through three long winters.

Pioneer Roots 15

If Bob and Pat Meyer were farming 150 years ago, they definitely would have been homesteaders. They perfectly exemplify Thomas Jefferson's legendary small family farmer: sturdy, generous, and hardworking.

■■■■

From the get-go, America was intended to be a country of small farms. The first European settlers worked as farmers, but instead of being the lowest laborers in the English feudal structure, farming in America developed into an opportunity—synonymous with self-sufficiency, prosperity, and, ultimately, the first American Dream.

Thomas Jefferson, the great "agrarianizer," believed that having an American economy based on family farms was essential to building a democracy. Richard Kirkendall, a historian at the University of Washington, writes, "In his [Jefferson's] view, such a farm conferred independence, since the people on it worked for themselves, not others, and it required self-reliance and hard work. Its most important product was the personality type required for a democracy ... "[1]

Jefferson, however, knew that in order to prosper, American farmers needed to produce a surplus that would establish them in the world market. Farmers in the emerging nation grew enough food for themselves while also producing huge amounts of tobacco, cotton, indigo, and grain for exportation. The Land Ordinance of 1785 was the first institutionalization of Jefferson's ideal. It "created simple procedures for the acquisition and distribution of public lands."[2]

By the beginning of the 1800s, American agriculture began to change rapidly. With the invention of the cotton gin and ox- and horse-drawn iron and steel plows, reapers, and threshing

Fields That Dream

machines, work became less manual and more mechanized. This technology encouraged an even faster westward expansion and fueled the passage of the Homestead Act of 1862. Historian Richard White explains, "The terms of the Homestead Act were generous and straightforward. The law provided 160 acres of free land to any settler [citizen or noncitizen] who paid a small filing fee and resided on and improved the land for five years."[3] Homesteaders flocked to the West, and by the late 1800s, Jefferson's vision for America had been fully realized—there were well over 2 million small family farms.

∎∎∎∎

While Bob and Pat are reminiscent of Jefferson's ideal homesteaders, they are on the other end of a very long story. Once the Homestead Act was passed, American pioneers expanded rapidly, like spilled milk over the table of a conquered country. Jefferson's vision of a nation of small farms was dependent upon the virtual erasure of the Native Americans.

Through a systematic genocide committed by wars, the intentional spread of disease, and destruction of natural habitat, an estimated 210,000 Indians were left of the 10 to 15 million people who were natives of America when Europe first made contact with the New World at the end of the fifteenth century.[4] This decimated population was then, through coercion, force, and broken treaties, corralled into reservations across the country.

Native American writer Elizabeth Cook-Lynn reminds us, "The invasion of North America by European peoples has been portrayed in history and literature as a benign movement directed by God, a movement of moral courage and physical endurance, a victory for all humanity."[5]

Pioneer Roots

This misguided celebration of our homesteading heritage has led us to a complex and often paradoxical American psychology of entitlement and denial, both about the continued social injustice toward minority communities and the rabid rate at which we gobble land—as though there was still a continent to be "settled." History books tell us only of the fortitude and ingenuity of our pioneer ancestors. While these characteristics may be true of individuals themselves, failing to mention the entire history of how we conquered the nation keeps us from learning from our past mistakes.

For example, Jefferson's ideal that a nation of small family farms formed the bedrock of democracy was built on a double standard. Jefferson himself was a plantation owner. His farm was worked by slaves. And the way to build this land of farms, a land of self-sufficient homesteads, was by annihilating the homes of others.

Similarly, we have to learn from the fact that there was no sense of conservation in the homesteading movement. Rather than rotate crops and leave fields fallow for a season, many homesteaders just moved on to a new piece of land once the old soil was depleted. It wasn't until land was harder to come by that farmers began sustainable practices. No wonder that a century and a half after the passage of the Homestead Act land is a scarcity. Without reflection on the truths of history, Americans have become caricatures of past pioneers, still insatiable, still going after the "American Dream." The new American homestead is the suburban house complete with a fenced manicured lawn. We are no longer a nation of self-sufficient family farmers, yet we have still homesteaded our way into a corner. Even on the westernmost edge of the continent, I have seen how suburbia is rapidly encroaching on Stoney Plains Farm.

Fields That Dream

Bob and I walk back outside to the greenhouse, where he picks out a baby Thai basil plant for me. Then we head over to the storage shed and he starts generously loading me up with potatoes to take home.

As he fills my bag, Bob says, "When we bought our first ten acres here, there was just one house up at the woods near the road, and the one house over there." He points toward a small house set back slightly from the road. I marvel at the differences I see and remember from my drive out to the farm all of the new houses going up like Cracker Jack boxes.

"It was all just open fields," says Bob, reminiscing while we walk to my car. "Now there are all these people living out here and commuting to Olympia. If farming is going to be viable, there has got to be affordable land, and it isn't. We've had half a dozen dairies close down in this area just in the last year. The big dairy right up the road now has houses on it. It used to be all pastures. I sit on the Agricultural Advisory Committee for the county, and we go through this every month. 'What can we do to keep farmers on the land or to make kids want to farm?'"

Ultimately, Bob is concerned about a sustainable future for the land. Looking back with hindsight that no homesteader from a hundred years ago had, he can see that the land is held in a delicate balance. He knows that farming is now one of the only ways left to preserve open space. How ironic that farming, which once destroyed the land and so many of its people, is now the key to preservation. It is like the saying in the land trust business: "Even the worst farm or ranch is better than the best subdivision."

Maybe this is why over the years he and Pat have bought

parcels of neighboring land so that their original ten acres has ballooned to twenty-five. Perhaps they are buying up property not to create surplus product to sell, as Jefferson envisioned, but to keep the land protected, the soil fertile. The Meyers have one of the best farms I have seen. Bob and Pat are gentle with their land, knowing that not just their survival depends on it. It is as though the Meyers are transforming the concept of homesteading from one of destruction to preservation—and just in time.

On my drive back to Seattle I think how ironic it is that our old notion of homesteading is still eating us alive. Americans have always wanted to conquer the country, first from wild land to homesteads, then from homesteads to cities and suburbs. We are like dogs chasing our tails in the eternal search for home. In the process, we are trampling and destroying the only home we have left.

In memory of Robert J. Meyer, 1938–2002.

Fields That Dream

Chapter Two: Good Girl/Earth Mama

I am sitting with Gretchen Hoyt as she eats an enchilada in the tangle of a kitchen that is under construction in rural northwest Washington. She is in her early fifties, dressed in jeans and polar fleece, and she has piercing yet very kind blue eyes. Gretchen is explaining to me how she went from being an early 1960s housewife to becoming a farmer. Or, as she had put it on the phone to me earlier this week, "I was one of the original back-to-the-landers. You know, a real earth mama."

Gretchen married young and moved with her husband from Washington to suburban San Diego where she became a "typical housewife."

"You know the kind," Gretchen says with a slightly ironic smile, "Monday you do the floors, Tuesday you wash, and Wednesday you iron. I always strove for the ultimate praise—being a good girl."

Gretchen was born just after World War II, when men returned from a war that had boosted the economy after the throes of the Great Depression. Women who had gone to work during the war were forced out of the factories and higher-paying jobs, and men were moving their families to the suburbs. The technological age had arrived, and the belief was that life was improving every day. There was a flood of new inventions to make homemaking easier, such as dishwashers, electric mixers, and an army of new cleaning agents.

The rapid advances in the technology of housewifery were one small part of this technological revolution, which applied to all areas of American life and particularly the agricultural industry. Farmer and writer Michael Ableman explains how the machinery of both world wars influenced farm practices: "Tank traction was applied to tractors; munitions factories and their nitrate reserves were converted to fertilizer production; and nerve gas stockpiles were used in pest control—the spirit of conquest turned toward the land."[1]

Many of these new chemicals and fertilizers were (and still are) petroleum based. For example, 2,4-D was developed during World War II as a chemical weapon. It was later transformed into the first man-made weed killer.[2] Similarly, DDT (dichloro-diphenyltrichloroethane) was initially used during the war to fumigate barracks and delouse troops.[3]

By 1945, there were 2.4 million tractors on American farms (a jump of 800,000 in just five years) and a "greater use of hybrid corn, mechanical corn and cotton pickers, combines, hay balers, peanut harvesters, milking machines, and chemicals."[4] The result: fewer people required to work the land, larger farms, a hugely increased petroleum dependency, and the rise of corporations such as John Deere (tractors), Pioneer (hybrid corn), and Monsanto (pesticides).

Fields That Dream

The food industry also began to explode with new inventions in the postwar era. Nutritionist Joan Gussow reports:

> At the end of World War II, a typical supermarket offered roughly 1,000 items, largely fresh, cured or canned commodities. In the post-war period, however ... new processing technologies developed for wartime feeding emergencies, suburbs full of housebound women working on a baby boom as well as a new advertising medium—television—stimulated an outpouring of manufactured foods. By the 1980's the number of food products had doubled, redoubled and redoubled again. New introductions in 1989 totaled over 12,000, that is 33 new products or product variations a day.[5]

The movement away from the land, small-scale agriculture, and from-scratch meals was a luxury for people who had grown up doing everything by hand. Ecologist Gary Paul Nabhan recalls visiting his mother and stepfather's house and being shocked to find only prepackaged food and scores of laborsaving devices such as electric meat-carving knives and bread machines.

> They no longer had to do stoop labor, or peel and dice vegetables by hand. They were finally freed from the menial chores that had been associated with food getting and food processing for more than ten thousand years. With these tools, with the Minute Rice that was premeasured into plastic bags, ready for boiling, and with Kraft Macaroni and Cheese Dinner poured out of

a box into the microwave, they had gained the leisure time that their own parents never knew.[6]

It was this empty "leisure" time that drove Gretchen and a whole generation of housewives to seek a deeper meaning in life, away from the predetermined world of *Father Knows Best*.

Gretchen explains in a quiet, somewhat proper voice that has an undertone of dry wit, "I felt like there had to be more to my life. I was bored and dissatisfied and depressed." She describes that the Vietnam War was escalating and young people were beginning to question authority in a way she never dreamed.

This observation gave Gretchen a glimpse of freedom, away from all of the "shoulds" of her childhood and marriage. She saw in the Vietnam protests a freedom of choice to go against authority—or what the "establishment" expected of a person. Gradually she realized that she wanted a different life. She took an enormous leap and left her husband, went back to school, babysat to make a living, and started her very first garden. Eventually she managed a gas station.

"It was when self-service gas stations were very new, and this was one of the first in the whole United States. We had this little building, and they would drive up and we would greet them. You had to say, 'Good morning, welcome to the Pit Stop.'"

She laughs heartily and then continues more seriously, "The owners wanted to employ women only, and wanted us to wear short skirts. But I let my employees wear anything so long as it was the right color."

Gretchen, originally from Puyallup, a town southeast of Seattle, eventually started making her way back up north.

Fields That Dream

One of the things that moved with her was the idea of the little garden she had in San Diego, which had taken root in her heart. That garden was the first soil she had ever planted, and she decided she wanted to make a life and living from growing things. Gretchen started out in Olympia farming communally on one of the many little fjords that jut out into the Puget Sound.

She remembers, "I didn't know anything about growing anything except I had this garden and it had chard and spinach that actually grew and I didn't have to hardly do anything!"

Eventually Gretchen decided to turn farming into a life. She met her future husband and business partner Ben through a mutual friend. Of Ben she says, "We became friends and eventually we moved on to wanting to grow together." She pauses and we both laugh at the double entendre.

■ ■ ■ ■

Alm Hill Gardens is in a town called Everson, about half an hour northeast of Bellingham and a stone's throw from the Canadian border. Everson is one of many small farming towns in this valley surrounded by the foothills of the Cascade Mountains. On the day I visit Gretchen, the valley is green, lush, and soggy with the last days of winter, and the cold air is a pungent combination of brush fires and cow manure. The land is dotted with dairy farms and farms that grow mainly berries. The population seems fairly diverse. Gretchen tells me that there are a lot of Dutch farmers in the area, and I see church signs in English, Korean, and Spanish.

The townships have been spreading out into the farmland for years. It used to be zoned forty acres to one house, but now it is only five. Most of Gretchen and Ben's neighbors are

Good Girl/Earth Mama 25

children of local farmers who inherited family land and put up a house, rather than keeping the farm active. Ben, who grew up here, used to feel like he knew everybody. Now he feels like he doesn't know anyone.

As I drive onto the property, I pass a huge field of tulips pushing their way through the rich brown soil. I make my way past several other fields before I reach Gretchen and Ben's house, which stands in the middle of a clearing. The house is dome shaped with some of the sides punched out to make more rooms, like a man with a potbelly that bursts the buttons on his shirt. Ben and his killer attack dog, a tiny dachshund with a mean bark, greet me. He is tall and lean and, with his glasses and distinguished receding hairline, looks like an economics professor. Ben tells me that the house has been under construction for the last twenty-five years. He says resignedly, "If we have the money, we don't have the time. If we have the time, we don't have the money."

Ben's father bought the property that is now Alm Hill Gardens when Ben was a teenager, and he is now fifty. Ben and Gretchen have thirty-two acres around the house. Originally they grew only berries.

Gretchen recalls, "It was real low-key. We went to another berry grower, dug up the plants from him. He was taking out a field and so we collected his posts and wire, and that's how we got started."

The first year, 1974, they planned to sell to the local cannery, but the prices were too low. Figuring there had to be a better way to make a living, Ben and Gretchen drove down to Pike Place Market in Seattle.

Gretchen recalls, "We would go with a carload of berries and come home with no berries and money in our hand. It

Fields That Dream

worked because growers used to have just one product. A lettuce grower might grow four kinds of lettuce, and that is all he sold for the entire growing season. It also worked because when we started selling at Pike Place, it was still local people coming to buy their local produce. Supermarkets cropped up after World War II, but a lot of people still had the habit of coming to the farmers' market. You don't find that so much anymore. Why go to a farmers' market when you can go to a grocery store, have free parking, and buy everything you need in one place?"

A number of years later they bought ten acres of river bottom about a mile away. They also lease five more acres from a neighbor. Ben and Gretchen now have a hefty operation of mostly tulips in the spring and lots of berries, bedding plants, and vegetables in the summer. They also have two part-time year-round employees and a crew of around a dozen full-time employees during the summer. All of the workers are from the state of Michoacán in Mexico.

She explains, "I have had many of the same people working with me for over seven years, and they have made me their family. We run our farm very differently than big factory farms where employees are expendable. My feeling is that my workers' jobs on the farm are the same as mine, equally important. It is just that theirs are different, and they are getting better at the jobs they do than I ever was at their jobs. We all specialize. I specialize in the management, where we are going to sell, what the price is going to be, how the packaging is, and all of that. They are making sure that they are picking the produce at the right stage of ripeness and keeping their hand soft on the product. Their job is so important that if there is any way I can make their lives easy and their workdays pleasant, I am going to."

Good Girl/Earth Mama

Gretchen explains that around 1986, amnesty was given to migrant farmworkers so that if they could prove that they had been working on farms for a certain number of years, they were given legal resident cards. All of the folks who work for her have legal residency, but they choose to live in Mexico during the off-season, where she has visited them a number of times.

"The economic situation down there is so hard that people are driven by desperation to look for work in the United States," she says. "If these folks can make enough up here to send money home so that the kids can eat, then they are going to do it. The real issue is that for many people, there is no choice but to immigrate illegally. I mean, these people are one generation away from mass starvation."

∎ ∎ ∎ ∎

We have moved into their living room and I am extremely cozy sitting by the woodstove, which Gretchen periodically stokes with big hunks of firewood while she tells me about how things weren't always this nice. "We spent several years financially pretty lean. We always had plenty to eat, but we didn't have power for four or five years. We had an outhouse when we had no plumbing. I mean, we have gone through some lean times. It is real character building. There is nothing like water coming out of a faucet after you haven't had that for a while. It is an absolute miracle!"

Especially when you are raising three sons and a daughter. The three boys are all grown and live out of state, and Gretchen and Ben's daughter is already sixteen. All of the children helped out on the farm in some way or another. Most of them sold at the farmers' markets, which Gretchen says helped them develop a real sense of confidence and ease with people.

Fields That Dream

Gretchen ponders, "You know, I worry about our youth today. The world is already overpopulated, so where is their place on this earth? I think the problem is that so many of these young people are lost. They don't want to do anything. Just drive in the car and go fast."

Gretchen and Ben are pillars in the northwest farming community and were one of the early supporters of the University District Farmers' Market, where they do an absolutely smashing business. I admire their farm's longevity, and I respect how they have made a successful business without seriously compromising their ideals.

But Gretchen and Ben have had to bargain a little. She explains, "We were organic way before there was a certification program. When we first decided not to be organic, it was a terrible blow. You know, we really believed in it."

Gretchen continues, "You see, when our daughter was born, I didn't really have time to do all of the work that's involved, the weeding mainly. And my husband said, 'I have other important things to do. I can't do it.' And that's when we decided to look into other things; namely, weed control. So, having been organic and being a commercial grower, I see that there are some benefits of being able to control problems. We are, however, very careful about sprays. The overall [picture]— what's going to happen with sprays, where they fit in, and what their long-term effects are—makes me feel very, very cautious. On the other hand, by not being certified organic we haven't placed ourselves in the position of potentially losing crops because we are limited in our resources."

She continues, "We do not spray for insects. We only spray

Good Girl/Earth Mama

to prevent weeds, and only when the plant we aim to protect is bare. If someone is going to place one of my crops in their mouth straight from my field, I don't spray it. Salad, peas, beans ... my vegetables are not sprayed. On my berry plants, I would only spray now, before there are any leaves. After the leaves come, the buds have to form, and then the berry comes from that. So six months before you get the berry, if you do something to the plant or the ground, it is a lot different than using something on the product that consumers are going to put in their mouths."

Gretchen is sounding world weary, as though she has justified their choices a million times before. She explains to me that she and Ben just got back from a farmers' market conference where, hanging out with almost all organic growers, she was not in a very popular position.

"There are the growers who say, 'I spray everything possible, just for the safety of it,'" Gretchen adds. "But we are just the opposite. I say, 'Let's be really careful about these things because they are chemicals that people don't normally eat.' So we are careful about what we use here, and we think about the alternatives and what would be the right choices."

▰▰▰▰

Gretchen and Ben's dilemma is not an unusual one. Needing to raise four children on the income of the farm alone brings up interesting challenges. Most American farmers don't rely on farm income at all, instead depending on government subsidies and off-the-farm jobs. Desmond Jolly, an agricultural economist, states that 90 percent of farmers' revenue comes from nonfarm income.[7]

The government subsidy practices that began shortly after

Fields That Dream

World War II are still in effect today. The large farms that emerged during the "Great American Agricultural Revolution" of the post–World War II era were dependent on technology and crop prices. These large farms focused on monoculture (producing mass quantities of one crop such as wheat, corn, rice, or soybeans). The mechanization of farms brought a huge amount of productivity, and there were large surpluses of crops that were used for school lunch programs or sent overseas as food relief. However, the crop surpluses brought the prices down so low that the farms were often not financially solvent.[8] To offset this danger, the government began providing agricultural subsidies to large-scale monoculture farming operations. Most often, farmers would use the money to augment the crop income, invest in more farm technology, and buy more land.

Critics believe that "Government subsidies are meant to help agribusiness, not foster an independent, efficient farming culture."[9] Indeed, very few, if any, small-scale farming operations, organic or conventional, receive government subsidies. A recent review of USDA records by the Associated Press found that "almost two-thirds of the $27 billion in federal farm subsidies doled out last year [2000] went to just 10 percent of America's farm owners, including multimillion-dollar corporations and government agencies … "[10]

Although owners of small-scale, sustainable farming operations were never content about not receiving government subsidies, for many years they were resigned to it; because being on the fringe of the market meant that while small organic farms didn't receive federal money, they were also not beholden to federal agricultural policies dictating their farming practices. However, with the astronomical growth of the organics market, the USDA determined that federal regulations needed to be

Good Girl/Earth Mama

implemented even though organic farmers would still not be eligible for farm subsidies. In the late 1980s, Congress passed the Organic Foods Production Act, which established a national standard for organic production and a method for informing consumers that products were organic. The National Organic Standards Board (NOSB) was formed to help the USDA write the regulation.[11]

In 2000, the USDA released a draft of the National Organic Standards for public comment. Nearly 300,000 people responded, particularly regarding issues such as whether or not to sanction genetically modified organisms (GMOs), irradiation, and sewage sludge as organic practices.[12] The USDA, astounded by the response, listened carefully to the public and modified the regulations accordingly. In October 2002, the National Organic Program was launched.

However, as Gretchen explains, the certification process allowing farmers to sell organic is long and arduous, and many feel weighed down by all of the government regulations and details. Of course, when a farmer is certified organic, she can sell her produce at a higher price. Although much of what Gretchen and Ben sell is essentially organic, I realize that they, like many farmers, have stayed uncertified to maintain freedom, and thus use mostly organic methods without the monetary reward.

Gretchen sighs slightly, "I think we have made a philosophical decision, and an economic one. You know, there aren't too many 'new-age' farmers who have been farming for twenty-seven years. We are successful. We did what we needed to do and we still do what we need to do to be successful."

Fields That Dream

We have gotten up to stretch our legs and have started to meander around the farm. Gretchen leads me into the greenhouse, which is magically warm on a cold day like today. It has that delicious clean dirt smell. There are a million bedding plant starts, blooming tulips, and, tucked in the back like fragrant jeweled treasures, pink peonies. We walk around the greenhouse and Gretchen says, "It seems like this is a great time to be a young farmer. You can have a couple of acres and you can sell food for six months out of the year and do pretty well. It hasn't always been like that. Finding the way to a market has been really hard. The grocery system is so built up that a little grower with a few cases of this or that has no place in the whole system at all. But I think the farmers' market scene is growing stronger. I am hoping that people will get into the habit of eating really fresh food. I am hoping that we will create enough of a movement that the markets will continue for a long time."

I understand exactly what she is saying. I have eaten peaches from the market that could stop traffic, and, as Gretchen points out, these are the kinds of experiences that make people happy and willing to stand in line to get fresh, tasty produce. Perhaps we are getting sick of our thirty-three new packaged foods a day and the disinfected, anonymous aisles of a supermarket. Going to a farmers' market is worlds away from ordering your groceries online, and people seem hungry for an authentic experience. A farmers' market is a way to get back to basics without romanticizing the past. Sustainable farmers today are translating elements of pre–World War II America into modern times, recognizing the advances of technology while still creating a healthy environment.

Gretchen continues, "We are going back to the simple

Good Girl/Earth Mama

values of flavor in our food that give life a different meaning. It is a simple thing, but a basic thing that can give you satisfaction."

We are laughing as we walk out of the greenhouse. I feel tremendous warmth for this woman who has moved such great distances in her life, from a picture-perfect housewife to an honest-to-goodness back-to-the-lander. I thank Gretchen for her time and remark on what an incredible story she has told me. She chuckles and says, " Being a housewife was my other life, another incarnation of me. I think it was all building and building and leading me toward this. Sometimes I look at my life, and it is not that different. I still cook the dinner and clean up the kitchen and do everything I always did. It is just out of choice now. I feel like I have a choice."

As I am driving away, I turn back to wave to Gretchen. She is standing underneath an ancient maple tree in the middle of one of her fields. The bare branches seem to reach for miles into the reddening evening of a late-winter sky. I see that she has truly found her place, and her roots are set as deeply here as the tree's.

Fields That Dream

Chapter Three: Natural Progression

About fifteen years ago, Jeff Miller made a leap of faith. In the middle of a very successful career as a chef, he quit his job, cashed in his life insurance policy, bought a whole bunch of lettuce seeds, and rode his motorcycle from California to Washington to start a farm.

This was a big change for him, especially since he had known he wanted to be a chef from the time he was five. By age fourteen he started working in his first restaurant, and after high school he applied to the Culinary Institute of America in Hyde Park, New York,

one of the top cooking schools in the country. After school, he returned to his hometown of Pittsburgh with his new wife, also a chef, and helped open a restaurant. Eventually, Jeff and his wife moved to California where he cooked in several four-star restaurants, including the very famous Stars. But after just a few years, Jeff decided he wanted to farm.

Jeff is explaining all of this while we sit at his

35

kitchen table drinking gigantic cups of green tea. He is hand-some—lean and muscular, with heavily lidded green eyes and rosy cheeks. His long brown hair is pulled back into a ponytail at the nape of his neck, he wears a beaded choker, and a tattoo snakes around his wrist.

I have a hard time seeing where and when Jeff's shift of focus occurred. All he says is, "It was just the natural progression for me, both creative outlets and both still being involved with food."

It all began when a friend of Jeff's went up to the Seattle area to visit his parents. When he came home to California, he told Jeff that no one was really growing specialty vegetables for the upscale restaurants in town.

Jeff recalls, "In San Francisco, farmers were bringing their specialty produce to the back doors of restaurants all over the place. A light just went on in my head. I was getting kind of burned out on cooking, and I was doing the more administrative aspect of it. One day I just said, 'I want to grow vegetables!'"

Before he dropped everything to pursue farming, Jeff did a bit of homework. He found a small quarter-acre farm in the middle of Berkeley. "There was a cyclone fence around it, and everything was grown in raised beds. All they grew were greens and baby lettuces, and I was fascinated by it. I used to go down there and hang out and watch them, and I knew that is what I wanted to do.

"So anyway," Jeff continues easily, "I ordered a bunch of lettuce seeds from a specialty seed catalog, I packed them in my backpack, and rode on up here."

Jeff modeled his farm after the Berkeley quarter acre, and for the first few years, he grew everything in raised beds and direct marketed to restaurants.

Fields That Dream

Jeff says with just a hint of pride in his voice, "You know, I didn't have a single account when I first started. And when my first lettuce became available I went to eight different restaurants in Seattle, they all loved it, and I started the business."

"You just went in and started talking to them?" I ask, impressed, and delighted with the idea of fresh produce coming to a restaurant right from the farmer.

Jeff laughs, "Well, I was a chef, so I knew how to do it. They always said, 'God, you are the first farmer who hasn't come in and tried to peddle his wares in the middle of lunch!' So I had an advantage. I knew how to talk to them; I knew what they wanted. So, it was relatively easy for me to get my foot in the door. I had eight accounts the first year, and then by the time I stopped delivering to restaurants a couple of years ago, we had about twenty-five different accounts."

■ ■ ■ ■

In this day and age, the idea of a chef and a farmer having direct contact is almost groundbreaking. One of the key ingredients of the post–World War II agricultural revolution was the rise of the interstate highway system. The highways brought about drastic changes in the landscape of America, what Americans eat, and the relationship between farmers and consumers.

During World War II, the Allied Forces had difficulty crippling German transportation. It was easy enough to drop bombs on railroad tracks and derail a train, but the autobahn was much sturdier. After the war was over, President Eisenhower traveled the autobahn and saw for himself how important highways were for national defense. Although World War II was over, America was now besieged by the Cold War and petrified of a nuclear attack from Communist Russia. Eisenhower felt

Natural Progression

sure that highways would be the perfect way to safeguard citizens if the need for massive evacuation arose.[1]

Many roads had already been constructed in the United States under previous presidential administrations, including the Alaska Highway, which was built during World War II. But Eisenhower's vision of interstate travel was greater than anything the country had previously witnessed, and it was actualized as quickly as possible.

The new highways transformed the auto and shipping industry, galvanizing the purchase of a record number of cars and trucks in the 1950s.[2] Moreover, shipping goods by truck increased 257 percent between 1955 and 1990.[3] Tom Lewis, author of *Divided Highways: Building the Interstate Highways, Transforming American Life*, writes, "These are the highways that bring grapes to Maine in the middle of winter."[4]

The construction of the nation's highways also made white America's postwar dream of affordable housing and communities for the swelling population of the baby boom a reality. During the "Great White Flight," the American middle class headed for the new suburbs, and it was the highways that took them there.

Suburb visionary William Levitt built the first of his many cookie-cutter housing developments on a potato field in Long Island, New York.[5] "Levittown" houses were available to white families only, and they spread like wildfire across the nation, cropping up in what used to be farmland and wilderness.

A land of farms and cities quickly transformed into a nation of cities and suburbs. Los Angeles led the way—400 people arrived in the county every day; 2,800 new faces a week; 146,000 a year. By 1960, more Americans would live in suburbs than cities. By 1980, eighteen of the nation's twenty-five largest cities

Fields That Dream

would decline in population, while the suburbs would swell by 60 million people, eighty-three percent of the nation's growth.[6]

Highways made it all possible. America's newly found mobility encouraged people to live in the suburbs where they would have to take the freeway to work and drive to the grocery store. All of a sudden, freeways weren't about national security, they were paved access to the American dream of getting a good job and owning a home.

Entrepreneurs such as Richard and Maurice McDonald capitalized on this emerging American "car culture."

> The McDonalds had understood an important new trend in American life: Americans were becoming ever more mobile and living farther from their work places than ever before. As they commuted considerable distances, they had less time and always seemed to be in a rush. Life in America was surging ahead and one of the main casualties was old-fashioned personal service. Their customers wanted to eat quickly.[7]

McDonald's and other fast-food chains boomed astronomically and eventually transformed every aspect of the food industry. How could fast-food not affect agriculture? McDonald's "is the nation's largest purchaser of beef, pork, and potatoes—and the second largest purchaser of chicken."[8] The postwar mechanization of agriculture fit in perfectly with the fast-food model, and the gap between farmers and consumers began to widen. Other restaurants capitalized on the efficiency of food production, ordering everything from produce, canned goods, and food mixes from huge grocers such as SYSCO, Rykoff, and Associated Grocers.

Natural Progression

By the early 1980s however, chefs including Alice Waters in Berkeley, Deborah Madison and Jeremiah Towers in San Francisco, and Charlie Trotter in Chicago wanted to bring high quality produce and locally raised meat back onto people's plates. They began to forge relationships with local farmers and foster a whole new movement that celebrated the connection between the growers and the diners. Soon their ideas traveled to cities across the country, and this is how Jeff Miller's Willie Green's Farm was born.

▪ ▪ ▪ ▪

For seven years, Jeff's farm was as mobile as he had been. When he arrived in Washington, he leased a couple of acres in Woodinville, a community east of Seattle. Then he moved to Monroe, a farming community, bastion of affordable housing, and a cute small town at heart, and leased a house with some acreage. Later he leased another plot of land in the Monroe Valley and lived in an apartment in town. Finally, about twelve years ago, the property that is now Willie Green's Organic Farm came up for sale.

Monroe is about forty-five minutes northeast of Seattle in the middle of an extremely fertile valley that stretches from the interstate to the foothills of the Cascade Mountains. The valley holds some of the Northwest's finest farmland, and an increasing number of suburbs. Large portions of Monroe's population follow in the previous generation's commuting footsteps. They get into their cars each morning to brave the minimum-forty-five-minute commute to a Boeing plant to the north or south to "The Eastside," a conglomeration of Seattle's sister cities that rests along the eastern shores of Lake Washington. The Eastside's cities are a home base for some of the world's largest computer companies, including Microsoft and many of the

Fields That Dream

other start-ups that have followed in its wake.

Monroe's main street looks like a soundstage for a movie that takes place in the 1940s. It is lined with cute storefronts and old clapboard buildings and has a quaint small–farm town feel. During the seven-minute drive from downtown Monroe to Jeff's place, one leaves behind the glaring white of new housing developments and passes a sign that says, "Entering Farming Country, Please Drive Slowly." And farming country it is, with its soggy spring fields, rows and rows of cover crop, ramshackle double-wide trailers, and little colonies of greenhouses sprouting out of the earth like mushrooms.

Jeff bought the Willie Green's land and built the property up from scratch.

"When this became available, it was just a piece of bare ground. We put in a septic system, dug a well, and put the house on."

The farm has twenty-nine acres total, sixteen of which are under cultivation. Jeff grows about four acres of assorted vegetables, four acres of berries, and eight acres of salad greens, of which he produces 1,100 pounds a week.

Jeff's cooking expertise comes in handy with his customers, who flock to his market stand to buy his impeccably grown greens and ask for advice on how to cook them. I am always asking him questions about how to cook rapini or the benefits of tatsoi over spinach. Jeff also has a million miscellaneous tips, such as how to keep green beans green while cooking them and the perfect amount of time to blanch a carrot.

Jeff has been selling at the University Market since the beginning. "The market is my favorite part of farming, because you have direct contact with the consumer, and it has a festive atmosphere. It is a long day. You are up at 4 A.M. loading the van and heading down there early to set up. But the interaction with the people is fun, and that is what I enjoy."

In addition to selling wholesale and at a few farmers' markets, Jeff also has a community supported agriculture (CSA) program where customers buy subscriptions and receive a big basket of produce every week for the duration of the season. The CSA is a good way for Jeff to set his roots down in the Monroe community. The current members live in the towns right around Jeff: Snohomish, Monroe, and Duvall.

Jeff muses about the Monroe community. "Although we have a good-sized CSA, it probably just scratches the surface of the people who would be interested. But it is a little harder in Monroe.

"Because there is a higher concentration of people in Seattle who are aware of their health, organic food, and natural healing, it has become like a mecca around the country for that kind of opportunity. I know there are people in Monroe who are like minded, but the concentration is just a lot more diluted, so it is harder to find people in this area interested in organics. On the whole it is easier for me to take my produce to the U-District Market and in five hours sell $2,000 worth of vegetables. There is an economic consideration, what is my time worth, and how can I maximize my time and make it profitable. A community connection is important to me, and at the market I can supply my community in a larger sense."

Maybe Jeff's love of feeding people that developed when he was a chef explains his passion for growing healthy and safe food. "Growing organically or nonorganically was never a question for me. It was always going to be organically."

Although Jeff didn't learn about organic agriculture until he was working at Stars, once he was exposed to it, the organic philosophy of sustainability made perfect sense to him.

Fields That Dream

Jeff reflects, "It resonated a need in me. If I am going to grow and live, I am going to try to have as little impact on the environment as possible, or perhaps even make it better when I leave from when I took it over. My connection to people in my community is being able to offer environmentally safe and sustainable food. Knowing that it is some of the finest produce people are going to buy. Period. With the added bonus of it being organic."

He continues, shrugging his ponytail off his shoulder, "I don't have an overwhelming need or desire to get into a grassroots or political movement immediately. I am so busy right now, I don't want to overextend myself, and spending time with my family is very important. I'm very happy with what I am doing right now. But it might be a natural progression for me in another ten years. Maybe then I won't even be farming anymore; I'll be an activist in some movement. But for now this is what I need to do."

I don't see how Jeff would have time to do anything else but farm. It seems to be all consuming during the season, even with the help of eight employees. "Most of them are high school kids," he explains, "and they come back every year."

Jeff laughs wryly, shaking his head, "Labor is a challenge on all farms. The economics of farming do not allow you to pay a lot, because people won't pay more for food. That is why the high school kids work out so well."

Jeff seems as if he would be a good boss, and potentially a good mentor if there were ever a kid out there who wanted to grow up and be a small-scale organic farmer. "I'm really laid back for one thing. But out in the field, it is like black and white for me. I am really intense, and I demand a lot from my employees. But I like to joke around and have a good time, too."

Natural Progression

I ask Jeff what his friends and family thinks of his farming. Jeff laughs and then shrugs. "Well, they all thought I was crazy when I first went into cooking, and then when I quit to go into farming they thought I was loo-loo. I don't think my dad will ever really understand why I am doing what I am doing. He is kind of a white-collar business guy. But my mom is really into it. She really respects what I am doing, and she loves to come and visit. She comes out every summer and helps on the farm. She loves getting her hands dirty."

That Jeff has been determined to make his farm succeed has helped him win respect not only from his family, but also from the farming community. Willie Green's has built an excellent reputation over the last fifteen years.

And he explains that it was pure determination that made him change his mind about cooking and head into farming. "I am very driven by nature, and I feel like I can do whatever I want to do if I put my mind to it. So I read a couple of books about farming, and that was it. There is really no formula that I can offer people. A lot of people get involved with farming, farm for a couple of years, and then just pack it up. It is more work than they thought. Either you do it or you don't. I just feel like if you want it bad enough, you will make it happen. And you are going to go through trials and tribulations and you will make lots of mistakes. Hopefully you will keep good records and just keep plugging along. By the third year that I was farming, I knew that I was going to make it. I knew there was a demand for my product."

Jeff checks his watch and tells me apologetically that he has to get back to work, so I head out on the road again. Jeff said to me that it was "freedom, plain and simple" that made him want to farm. In spite of myself, I feel a similar burst of freedom and

Fields That Dream

independence when I swing back onto the highway, heading south toward Seattle.

Unlike Jeff's natural progression from chef to farmer, our country has made a complicated transition from a nation of bumpy roads and farmland to supersmooth highways and acres of strip malls. While these roads have brought us closer by making travel more accessible and affordable than ever before, they have distanced us from one of the very basics of life: eating food grown in our own communities.

But I think Jeff and farmers like him are reclaiming the roads to keep produce close to home, usurping the highway right out from under the SYSCO eighteen-wheelers. His motor-cycle journey to Washington was the first wave of a rebellion to recover land destined for suburbia. Maybe it was Jeff's destiny to swoop up that piece of earth. He said to me, "The need to farm was within me, and I was lucky enough to discover it, or it discovered me. I don't know which it was. I think it found me, and I am fortunate." But we are the fortunate ones, now that the road is coming full circle.

Chapter Four: Moving Easy in

It is a crisp, gray Saturday in the middle of winter. The sky is blanketed with clouds so high that the magnificent snow-covered Olympic Mountains west of Seattle are in perfect view. When I drive off the ferry onto Bainbridge Island, the Seattle skyline looks like a picture postcard in my rearview mirror. I am headed to Port Madison Farm, which is about a ten-minute drive from the ferry landing.

Steve and Beverly Phillips operate one of the only goat dairies in the entire state of Washington among a handful of farms on an island that is largely residential.

Had the Phillipses started their farm on Bainbridge Island in the 1920s or '30s, they wouldn't be such an anomaly. Before World War II, Bainbridge Island was largely agricultural, farmed by many Japanese-American families. With the internment, however, the Japanese-Americans were forced off their land and into camps while their farms were taken over by suburbs.

Today Bainbridge Island, thirty-five minutes across Puget Sound from Seattle, is a strange conglomeration of quaint rural island and upper-middle-class suburb. There are strip malls, fast-food restaurants, and gas stations, garish three-story water-front homes and rambling split-level ranch houses. But there are spots where the roads become narrower and the sidewalks disappear, trees canopying thoroughfares edged with black-berry brambles. The only company seems to be the large black crows that perch on branches in the rambling woods.

I eventually find Port Madison Farm at the end of a long, unmarked gravel driveway off a quiet residential road. A two-story wood-sided, weather-aged home, beautiful in its simplic-ity, is set back on the edge of a large clearing with an expanse of woods surrounding it. Scattered around the clearing are pastures and two barns, one huge with a milking parlor attached in the rear. Dense woods surround the area, and the air is laden with the sharp fragrance of pine needles.

I am not the only visitor to the farm this morning. A mini-van with its motor still running sits in the driveway before me. Two men are talking, one of whom I can only assume is Steve Phillips, since he is wearing work boots, a bright red down vest, and is holding a wrench. The driver of the minivan is dressed in his Saturday casuals, and clouds of both men's breath billow into the cold air as they laugh and talk. A little girl sleeps in her car seat on the passenger side of the minivan, and I can hear her father saying, "She was so excited about coming here. All she talked about was going to the farm to see the goats, but then she conked out. You know how kids and cars are." He walks around to the back of the car to retrieve a number of empty glass milk bottles.

"Thanks so much, Steve. We are really glad you are doing

Fields That Dream

what you are doing. We really appreciate being able to drop by and see where all the magic happens." He gets back into his car and calls, "See you soon!"

As the minivan pulls away, Steve notices I am there, puts down his wrench to shake my hand, and calls to Beverly. Steve has dark hair that is flecked with gray, brown eyes, and deep, rich olive skin. Although he is kind, he seems slightly wary of me. Beverly emerges from around the large storage shed, brushing hay from her sweater. She is absolutely lovely, with pure silver hair and rosy flushed cheeks.

We stand around chatting while Steve tries to fix the muddy orange rototiller. After a few minutes of talking about the farm, Steve's initial wariness begins to give way to a genuine enthusiasm. He explains that he was a furniture designer for years, but says with frustration that "The business was very impersonal. Once my customers got their furniture, they didn't care whether I lived or died. But that guy who was just here, he thanked me for the milk. You just don't get that kind of response in the furniture industry. That is one of the reasons why I like farming."

Steve continues, "You know, this farm can be enormously gratifying. The work is unrelenting, and I have to make my peace with the fact that I'm never really going to be on top of it. But I do like that we built everything here. Beverly even framed the house by herself. I like that we have virtually total control of what goes on here. But it is a double-edged sword, because when something goes wrong, there is no one to curse at but each other!" They both start to laugh, and I admire the obvious delight they take in each other.

Beverly made her own journey into farming. Although she was a computer programmer for the telephone company for twenty-two years, she declares that she has always been a farmer at heart. Beverly grew up in rural Maine and then moved with her first husband to Alabama, where she lived for ten years. After her divorce, she and her eleven-year-old son took a summer to wander the country to find the place they wanted to live. Eventually they landed on Bainbridge Island.

Beverly thought that she would start farming as soon as she moved to the island, but she couldn't even buy a building lot for the income from her whole house in Alabama.

She exclaims, "The difference in expenses was incredible. I had to go back to work to support my child, but eventually was able to buy this property in 1984. So, I guess I'm doing now what I started to do when I was thirty, which is be a farmer!"

Beverly and Steve met in 1985 and carried on a long-distance relationship before he decided to move from Berkeley, California, to Bainbridge Island. They began growing organic vegetables for a community supported agriculture program, but it didn't last long when they found that they were really just growing their own food and selling the surplus. They decided to start a dairy, built the barn in 1990, and were licensed in 1992. For that whole time Beverly was still working for the phone company, telecommuting full time until August of 1998.

They found that from a marketing standpoint, the Seattle area was the perfect place to offer fresh goats' milk and gourmet goat cheese. Steve says, "We are in the middle of 4 million people who are aware, conscious, and affluent. But, from a production standpoint, it is hellacious."

Port Madison Farm is the only grade-A dairy in Kitsap County, cow or goat. The closest dairy is in Sequim, at the northern end of the Olympic Peninsula. The nearest dairy veterinarian is in Mount Vernon, a ferry ride and an hour and a half drive away, so they do virtually all their own veterinary work.

50 Fields That Dream

Getting hay delivered is a whole other issue for Steve and Beverly. There is no one on the island that can teach Steve how to grow it, and his tractor is broken down anyway. He quips, "Auto mechanics don't make house calls!" As a result, they have 150 tons of hay, fifteen tons of straw, and between thirty and forty tons of grain trucked all the way from eastern Washington.

Steve explains the lack of resources on Bainbridge: "This is an island of doctors and bank executives. I had a friend of mine's son come over to muck out the barn, you know, a high school student. ... He drives up in a $40,000 sports car to clean out my barn." He pauses and lets out a howl of laughter. "He was working because his dad told him it was good for his soul."

Steve laughs again, this time with a tinge of irony, and then grows serious. "Every penny we have is tied up in this dairy. Our goats are worth about $300 to $600 each, and it would only take one day to ruin them all."

■ ■ ■ ■

The Phillipses run an old-fashioned family dairy farm, the kind of place where the farmers literally depend on the lives of their animals for their own survival. In the early 1950s there were more than 2 million small dairy farms like Steve and Beverly's. By 1992, there were only 155,000 dairies, with many thousands of cows on each.[1] As one would guess, corporations rather than families own these new "factory farms," and the care of the livestock has plummeted to grave conditions.

Writer Erik Marcus further explains:

Animal husbandry reflects the farming ethic of the first half of the 1900s, when farmers strove to provide an ideal environment for their animals. Farmers pro-

Moving Easy in Harness

vided such care not necessarily out of generosity, but because it was the only way to make the animals thrive sufficiently to generate profits. Today's farmers have no need for this ethic. Fifty years of animal science have developed an arsenal of drugs, hormones, systematic mutilation techniques, and specifically bred farm animals. Together, these developments let farmers raise animals more profitably—but under harsh and crowded conditions that would have killed earlier farm animals.[2]

This drastic change in the number of family-owned farms and the conditions of the present corporate-owned farms are explained by the frenzy of mass production that permeated the air in postwar America. R. W. Trullinger wrote in the USDA's *Yearbook of Agriculture 1950-1951*:

Everything is geared for mass production on a big-business scale, and it would be difficult to believe that the agricultural industry of today will voluntarily revert to last-generation methods and practices. Efficiency is of paramount importance in farming now ... In wartime, when technology moves faster, efficiency becomes even more vital.[3]

This idea of efficiency was also paramount to the homogenization of culture that developed in the '50s, manifested in part by the rapid rise of fast-food chains that appeared after interstate highway construction. People began to want the same product everywhere they traveled, whether it be rural Alabama or big-city Chicago. To meet the consumer's desire for mass-

Fields That Dream

produced food, farms had to simulate factories. On factory farms, the animals are housed in huge warehouses where they can barely move and never see the light of day. There are hundreds, sometimes thousands of animals for every worker, and livestock is looked at only as a dairy/meat-producing machine.

This leap to large-scale meat and dairy production has not only affected the lives of animals and farmers, but the land itself. Meat production has quadrupled in the last fifty years, and there are now 20 billion livestock on the planet.[4] This growth has affected the environment, from water supply to deforestation to pollution. According to the Water Education Foundation, it takes 2,464 gallons of water to produce one pound of California beef.[5] Manure from livestock should be able to be used as fertilizer, but with such "huge numbers of animals concentrated in feedlots and confinement buildings, there is no economically feasible way to return the animals' wastes to the land."[6] Instead, the millions of gallons of manure are stored in open "lagoons" that sometimes leak and overflow. Furthermore, aside from the topsoil erosion in the pastureland of America, Mexican and Central American rainforests have been decimated for cattle grazing—meat which is then imported to the United States.[7]

■ ■ ■ ■

When Beverly leads me in to see the animals, leaving Steve working away on his rototiller, I find that Port Madison Farm is the exact opposite of a factory. Entering the barn, we find delighted chaos as soon as the goats see us. I am met with the sweet smell of alfalfa and the unmistakable scent of goats. A bleating chorus of goats tumbles over themselves to nuzzle up to their visitors. They are gregarious, affectionate, inquisitive, and adorable.

Above the din, Beverly says, "If you have been around children, you know they have a mad cry, a hungry cry, a scared cry, and a hurt cry. This is a 'pay attention to me' cry!"

The winter barn is enormous and comfortably houses 150 to 200 goats. During the summer there is a smaller barn where the goats go when they are not in the pastures. They are kept by the dozen in roomy, straw-lined pens.

Beverly explains, "Mostly we like to keep the goats in groups of twelve. That way they stay together their whole lives. They have a group identity. If we tried to mix the girls in this pen with the girls in that pen, it would be gang war! They are really bonded to their pen mates, and they don't like anybody else in there."

Beverly beams as she leads me around introducing everyone. "This is Eloise, my first goat. She is thirteen," she says proudly. "Everyone here is either Eloise's granddaughter, great-granddaughter, or even great-great-granddaughter. I named her after the girl in the children's book *Eloise*; you know, the one who lives on top of the Plaza Hotel in New York."

In the pen with her is Mimi, Eloise's first daughter. Beverly laughs, "She named herself because when I used to go out and feed the goats she would go, 'me! me!'"

I am nudged from behind and nearly topple over. Beverly laughs at the small brown and white spotted goat that is as playful as a puppy. "Oh, that is Pippin. She lives in the aisle. She prefers to be by herself. She goes back to her pen at night to sleep, but during the day she jumps out and wanders around."

I ask jokingly, "So with all these ladies, how many men do you have?"

"One man, eight bucks," Beverly replies laughing, as she leads me over to a pen where the male goats are kept. I tell

Fields That Dream

her that I am under the impression that bucks are kind of aggressive and need to be kept separate from the other animals.

"No, the bucks are really just darling. They are all sweetie pies. They are all bottle fed, so they think I'm mama," says Beverly as she leans over a wooden pen to pet a particularly large goat who jumps up onto the rail and extends his front legs as though to give her a hug. "This is Lazarus. He almost died when he was a baby. To this day I don't know what it was, but he was only two weeks old and he stopped eating. It may have been an intestinal blockage, and he went down to skin and bones. I was sure I was going to lose him, and then for some reason he had a spontaneous recovery." Beverly coos as she nuzzles him, "We think it's the power of prayer, don't we, Lazarus?"

Everyone in the barn is special, and everyone has a name. We walk around for a while longer while Beverly introduces me to many more goats including Maryanne, Eleanor, Jane, and Elizabeth, named after characters in Jane Austen novels she was reading when they were born.

My first inclination is to pet the goats who are excitedly clambering for attention, but Beverly asks me to please not touch them for fear of outside germs that would endanger their health and safety. As Beverly and I walk out of the barn to rejoin Steve, she spells out their protective measures.

The Phillips do not use the stockpile of antibiotics and other drugs that large farms saturate their livestock with, in turn jeopardizing the health of the animal and wearing down the effectiveness of medicinal antibiotics on human illnesses. Their first concern is for the health of their goats, and they are very cautious. For example, their herd is entirely homegrown to eliminate the possibility of outside contaminants.

Moving Easy in Harness

Limiting antibiotic use is unusual in the world of dairy farming, where every year nearly 25 million pounds of antibiotics are administered to livestock for purposes other than treating disease.[8] The Phillipses don't use hormones to stimulate growth or milk production either. Hormones such as progestin, estrogen, and androgens, which have been linked to cancer and reproductive disorders, get into the human system through meat and in the waterways through manure dumping.[9]

To ensure the safety of the goats, the Phillipses have even had to sacrifice being certified organic because they are not able to get good enough quality organic grain and hay. Steve says, "We simply buy the best we can get, and sometimes it is organic, and sometimes it isn't. We used to buy organic hay, but then we stopped buying it because it seemed like every bale contained cowpies. The last thing we want to do is feed cow manure to our goats, because we try and avoid bringing in outside germs and diseases. We decided that the health of our goats and the people that drink our milk was more important than the label of organic. Except for the fact that the hay isn't always certified organic, we follow the organic principles."

This consideration and effort for the welfare and safety of their animals and the quality of their product sets the Phillipses apart from the standard American dairy farmer. By keeping their farm focused on serving the local community rather than mass production for a larger market, Steve and Beverly have enabled Port Madison Farm to be an ecologically sustainable operation that treats the land and livestock with respect.

We reach the house and I settle into an exquisitely carved wooden chair that Steve made. The interior of the house is

Fields That Dream

beautiful, with large windows and gleaming hardwood floors. Beverly is sitting on an upturned milk crate cradling a baby goat. There are a number of large, plushly lined crates near the heating vent that contain about seven newborn goats whom Beverly is in the process of feeding.

Beverly says confidently, "We run our dairy in a very different way from anyone else that I know. It is not just an economic decision. A lot of the way we do things has to do with our commitment to the animals. There are things that other dairies do that we don't because of our commitment to the animals. For instance, raising the babies. Most dairies leave the babies with the mothers for five days to clear out the colostrum, the most nutritious of the breast milk, to save time, and then they take them away. By that time the mother and the baby have bonded, and it is a terribly wrenching thing for both of them."

Steve adds, "Goats are a herd animal, and herd animals have really strong ties."

"Right," Beverly continues vehemently. "Rather than put them through that trauma, we take them away at birth. I mean, the goats bond to me. Every goat out there literally thinks she gave birth to me because I am right behind her. I wrap the baby up in a towel and she turns around to see whom she gave birth to, and she sees me. She licks me, and talks to me, and she thinks I'm her baby. Every week someone calls and says, 'How do you get your milk to taste so good?' The reason why our milk tastes so good is because there is love in that milk."

I must look slightly puzzled because Steve turns to me and gives a scientific explanation: "One thing about goat's milk that is different is that it picks up flavors more easily than cow's milk because of the different molecular structure, and goats

Moving Easy in Harness 57

are very easily excited. Adrenaline has a very strong and very unpleasant flavor, and that is what gets into most goat milk and cheese because they are being milked by people who don't know the animals."

Beverly interjects, "But for us there is a relationship involved here. Since every goat thinks she gave birth to me, she wants me to milk her. It wouldn't do for a stranger to come in and try to touch her. She would freak out."

Steve adds, "A stranger in the milking parlor makes them nervous, even though the actual milking is done by and large by our milking machines."

"But it is me washing her, it is my voice she hears. We are their babies," Beverly emphasizes.

Their commitment to the goats keeps Steve and Beverly intensely tied to the farm. Some of their neighbors have kindly offered to help out, but Steve jokes semi-seriously, "It is not like coming over to feed the dog and water the plants." He continues, "You have to milk at the exact time every day because if you make the goat wait, her udders will get tight, and her automatic nervous system will think it's making too much milk and try and shut down a bit. In just one time, the milking procedure can be thrown entirely off. Going away is simply not an option, because there is no one trained to care for the goats. There are times when it is unrelenting. There are times when I think it would be good for us to take a break. But it is a trade-off. If we were to get an employee trained for milking, we would have to have 50 percent more goats to make it worth our while financially."

They do have one employee, who is like Port Madison's ambassador to the world. Suzy goes out and sells at the farmers' market, where there is always a line to buy Steve and Beverly's

Fields That Dream

luxurious chèvre and other cheeses. She also delivers cheese and milk to a number of stores throughout the Puget Sound.

There are sacrifices of other kinds as well. Steve's parents are getting old, and when one of them dies, he feels that he won't be able to go back to the funeral.

With frustration he says, "I wasn't able to go to my daughter's high school graduation." He pauses, "I don't expect that either of our children will want to have anything to do with the farm. That is tough. It is real tough."

Beverly sighs and she places a just-fed, sleeping baby goat back into its pen and reaches for another. "Sometimes I think we are crazy. I mean, it has been ten years since we have slept past six in the morning. We haven't been able to go to dinner or see a movie. I mean, it was a pretty drastic lifestyle change. Anytime we leave the farm, which is rare, even if we are gone for just two hours, we are anxious to get back. We can't go somewhere all day. I mean, everybody needs lunch!"

For all of this hard work, Steve and Beverly look well rested and absolutely radiant. I tell them that it must be because they love their work.

Steve chuckles, "Yeah, I think we are moving easy in harness." He turns to Beverly, "Isn't that the term?"

"Oh yeah," Beverly replies. "Robert Frost's definition of freedom has always been my favorite: 'Moving easy in harness.' We are free because we are harnessed to what we want to be harnessed to."

Beverly continues, "We are here for the long haul. It gives me a different feeling about my community because I care more about what happens here. And I think the community values us."

Steve adds, "On the physical level, farms tend to be water recharge areas, they keep down population density, and driving past a farm gives you a different feeling than driving past a development. And on another level, if we sell lousy milk our friends and neighbors will know, and they know where to find us!"

I ask about the other harsh realities of the dairy industry that they face, namely that there is no real use for the surplus males that are born.

Beverly gets a stoic look on her face and says, "We don't sell our goats. Most dairies, when they are done with a goat, send her down the road. They go for meat or to auction. We don't do that. Our goats have tenure. Obviously we can't keep them their whole lives, but if they are going to die, they are going to die here, among their friends, and with a smile on their face. They are not going to know what hit them, and they are not going to suffer or even worry."

Steve says, "That is the hardest thing that I do, putting down the goats. They have been doing everything they can for me, so how could I ship them off to an auction where they will be horribly mistreated or sent to a slaughterhouse?"

"We used to try and place them with families as pets. Like the boys that are born that we can't use. But they will bring the goat back, or we will get a call in the middle of the night, 'The dog just chewed up the goat,' or ' The goat just ate a rhododendron, what do we do?'" says Beverly.

"Goats are not a suburban animal. They are farm animals, and they need to be kept in a farm situation behind farm-grade fences. Selling them is a prescription for disaster, so we don't sell them anymore. The boys that we don't keep for breeding are put down at birth. It is the kindest alternative."

Steve says sadly, "That is the downside of farming. It is tough. You are making life and death decisions every day. Animals should not just be a commodity."

As I wait in the ferry line heading back to Seattle, I think about how different the American landscape would be if it was still dotted with dairy and meat farms the way it was fifty years ago. Granted, Americans would have to be more mindful of the immense quantities of meat and cheese that we consume, but, with our high levels of obesity and heart disease, that might be a good thing. In exchange, we would be getting meat and dairy from the neighborhood farm. The perils of mass production, such as meat-borne·diseases, environmental degradation, and the mistreatment of animals, could be all but eliminated.

But I also see that the life of a small dairy farmer is filled with long hours and intense commitment. I am struck by Steve and Beverly's intelligence and overwhelming dedication to their goats and customers. And yet, the responsibility they hold over the health, and the lives and deaths of their animals is staggering. In my mind, I can hear the weight behind Beverly's words as she said, "Our goats are family. This is not goat 246, this is Eloise."

Chapter Five: Backyard

"So, I heard something interesting on NPR yesterday," John Huschle says as he takes off his stocking cap and precariously balances a small tray on which rests a teapot and cup. He sets the tray on our table and then plops into the chair across from me. "It was about Machiavellianism. They were basically saying that in the ways of capitalism, the toughest tend to succeed the most. I was thinking that there are a lot of things in farming that lend themselves to capitalism, and a lot that don't. But parts of the philosophy seem true. The tougher you are, the further you go."

John pauses, runs a hand through his mass of short, thick, and somewhat wild brown hair. "But the age I am at now is pivotal; I am going to be thirty in June. And I got to the point where I had to stop and reevaluate and start a different career if I needed to. I

realized that in farming I needed to have less control and more involvement."

He is referring to a difficult business decision he made a couple of years ago. John had hooked up with his best friend from high school just after college graduation to start a farm. He and his friend apprenticed on a farm outside of Minneapolis, their hometown, for a summer, and then headed out West.

John blows on his tea, takes a tentative sip, and then says, "There is a lot more opportunity out here. In the Midwest there are a lot of farmers, period, and a lot of farmers that transitioned to organic a long time ago. Those farmers have the market locked up tight because there are not as many people out there interested in organics as there are out here. It seems like with the disposable income and the progressive attitudes, Seattle just eats up whatever I grow."

John and his friend moved out to Seattle in 1995 and leased land immediately. John confirms that the leap into business was intimidating, but they managed to implement all they had learned during their internship and made it through their first season. John and his friend farmed together for another season and then decided to go their separate ways.

He explains, "Basically, I had been wanting to stay small. The farm we started is still small in the broad terms of organic agriculture, but I needed two things that our farm didn't offer: I wanted to have fewer employees and I wanted to regain control of production. We had ten or twelve people working for us, and I was just running around telling people what to do. I missed the actual hands-on work of farming."

As it stands, John had decided to leave one of the most financially promising new farms in the area to branch out on his own. His new farm, Nature's Last Stand, is located near the small town of Carnation, one of the last remaining farm belts in King County. Carnation, only about forty minutes from Seattle, is in the Snoqualmie Valley, which unfolds into a patch-work of lush bottomland formed by the ancient glaciers of the Cascade Mountains, which, on a clear day, are visible directly to the east.

John leases twenty acres of land, five of which are in pro-duction. The rest of the acreage is in orchards and being cover-cropped for rotational use. He lives on the property in a yurt, which he just put up, and a crew of two seasonal employees live in travel trailers during the summer.

He says passionately, "To me, quality is everything. At the other farm we did a lot of wholesaling, and I never want to sell wholesale again in my entire life!"

He chuckles at his own vehemence and then continues, "I really want to sell my produce directly. In order to go direct you have to have an emphasis on quality. I felt that it was harder to focus on quality when you have a ton of people doing the work. If you are out there with ten people and you can do the task better than they can, you should be doing it."

■ ■ ■ ■

Obviously, John is not your average Machiavellian capitalist, but he is still a man of the new millennium. He proved it when, after a monthlong game of phone tag, I finally reach him via his cell phone. He laughed and said, "Yup, I have finally become one of those people I hate who talk on their cell phones at the grocery store!"

Now, a few days later, on a rainy February evening, we are meeting at the Teahouse Kuan Yin, a serene little spot in the Wallingford neighborhood. The teahouse is one of those quintessentially "anti-Seattle" Seattle establishments. A sign firmly states, in calligraphic lettering, that "espresso and cell phones are prohibited." I get my favorite World Peace herbal blend tea, and John experiments with some green tea variety that I cannot pronounce. We sit at a little table for two in front of a huge fish tank, and there is plinky harp music playing in the background.

I have been trying to get together with John since November. With John's winter growing season going strong, the tumult of the World Trade Organization (WTO) meeting, and the holidays, the months just sort of slipped past.

We begin to talk about the WTO protests, and John says carefully, "Honestly, I got a bit frustrated with all the protests at WTO. I think protest activism sometimes is so focused on people pointing fingers, shaking their fists at someone else or something else, and they are never stopping to realize their own part in it all. I mean *our* own part in all of it."

John's comment hits home. The WTO protests were a couple of months ago, and I am still mulling over all that I learned—thinking about how although I may have been a protestor, I am still a "consumer" and share in the responsibility of the international economy. I was at the World Trade Organization demonstration in downtown Seattle on November 30, 1999, where more than 50,000 people converged on the streets to protest the organization's policies. The sheer energy of so many thousands of like-minded yet diverse people was exhilarating. There were young radicals, Mexican farmworkers, "middle-American" steel- and autoworkers, scientists, environmentalists, religious leaders, social justice activists, farmers, and laborers from all over the world.

▨ ▨ ▨ ▨

The history of the World Trade Organization goes back to the catalytic time just after World War II. The postwar industrialization of agriculture was one piece of the massive movement to promote corporate interests in the world economy. Toward the end of the war, members of the United Nations met in New Hampshire at the United Nations Monetary and Financial Conference, a meeting more commonly known as Bretton Woods. Here the World Bank and its financial backer, the International Monetary Fund (IMF), were born.[1] The World Bank finances the building of roads, agricultural projects, factories, dams, and power plants in Third World countries, among other ventures. These projects are most often initiated and completed by First World corporations and have become the infrastructure for international trade.[2]

The General Agreement on Tariffs and Trade (GATT), established in 1947, was a natural consequence of the World Bank and the IMF. GATT sought to bring the global economy under control after the chaos of the Great Depression and World War II and create regulations for the projects financed by the World Bank by "building a framework of rules to oversee international trade and taxation."[3]

After nearly forty years, trade officials began to aspire to expand GATT's powers, and in 1986, at the beginning of the Uruguay Round negotiations, they "started working to establish a permanent international body that would have the power to enforce GATT rules through monetary penalties."[4] The negotiations lasted until 1994, and the WTO was established on January 1, 1995.

Both the GATT and WTO have been highly controversial since their respective inceptions, and the lines of opposition are not clearly drawn along liberal and conservative factions. There

Backyard Homestead

has been tremendous opposition to international trade policies by environmentalists, trade unions, small business owners, and human rights advocates and strong support by influential corporations such as Hewlett Packard, Boeing, Cargill, and Monsanto.

The WTO includes 146 participating countries, accounting for more than 97 percent of world trade.[5] Where GATT originally focused on the trading of goods such as agriculture and textiles, the WTO has expanded to cover services like telecommunications and transportation, finance, and intellectual property including copyrights, patents, industrial design, and more.[6]

Similar to the United Nations, each member country's government votes to join the WTO, and it is governed by non-elected officials who are appointed by their respective governments. In late 1994, the WTO legislation was rushed through Congress by "fast track" rules where "Congress would have no power to modify any part of the agreement, and Congress was forced to vote up-or-down on the deal within 90 days."[7] Despite fierce opposition by both liberals and conservatives, few members of Congress actually read or studied the 22,000-page WTO agreement, and the decision to join the WTO was made without fanfare or publicity.

Supporters and opponents alike agree that the WTO has an enormous amount of power. For example, if at a WTO meeting members feel that a national or local law is a barrier to trade, then the WTO can impose trade sanctions aimed at changing the law to meet WTO standards. The Northern Plains Resource Council reports that, "To date, every national public health or environmental law (including U.S. laws) challenged under WTO rules has been ruled an illegal barrier to trade."[8]

Essentially, this means that laws surrounding workers' rights, health, and environmental preservation can be disre-

Fields That Dream

garded as trade barriers. The WTO also heavily affects agriculture. Activist Peter Rosset explains that the WTO is:

> Forcing countries to remove barriers to agricultural imports, driving down farm prices, and bankrupting local farmers. Stopping governments from subsidizing their small and family farmers, eliminating social safety nets. Pushing countries to compete in the global economy based on low wages, union busting, cheap production costs, and weak environmental laws. Extending U.S.-style patents over life forms worldwide, allowing corporations to commodify genetic plant resources that were once the heritage of local communities. [And] Making it difficult for countries to restrict the import of GE [genetically engineered] foods and seeds.[9]

There was a lot on the table for the meeting in Seattle; namely, discussions of farm subsidies, anti-dumping campaigns, and intellectual property rights. In a speech at the large labor rally, Dr. Vandana Shiva, a scholar and activist from India, connected her opposition to the WTO with the patterns of consumption so evident in a world of strip malls and fast food:

> The trade policy shaped by the WTO's free-trade paradigm is pushing the world to hyper-consumerism while erasing the evidence of destructive consumption patterns. The right to consume anything from anywhere with total ignorance of the true costs has been redefined as the new concept of freedom. However, such freedom is based on taking away the freedom to

survive with integrity, dignity, and well-being from millions of people and millions of species.[10]

I was stunned. In that massive crowd of protesters, I stood there and truly realized for the first time that I am part of the problem. In America, where most of us have such wealth compared to many countries, our rate of consumption, which we see as normal, is significantly higher than the rest of the world. Just a trip to an American grocery store is an example of this skewed level of consumption: we have an entire aisle of breakfast cereals alone to choose from. Although I had begun looking at the agricultural system in this country, I had no idea about the global ramifications of our agricultural practices.

I began to feel extremely powerless and confused. Why were there 50,000 of us protesting if we are all part of the problem? Perhaps all of the other people who converged on the streets of Seattle had similar feelings of powerlessness. We do not know how to eat, clothe ourselves, go to our jobs every day without harming someone inadvertently through what we buy.

I wondered how we, the consumers of the world, could get this far. It is as though we have created a "global homestead," taking the idea of Manifest Destiny a bit too literally. Historian Frederick Merk explains Manifest Destiny as " ... expansion, prearranged by Heaven, over an area not clearly defined. In some minds it meant expansion over the region to the Pacific; in others, over the North American continent; in others, over the hemisphere."[11]

It appears as though the world has been conquered, and Manifest Destiny is the patron saint.

I walked away from the protests with my eyes opened. Since then I have been trying to trace back to the origins of

Fields That Dream

what I consume, attempting to be aware of my own part in
fueling the fire of globalization. It is an overwhelming process,
and I feel entirely confused. I have been trying to figure out
how I can protest something when I am part of the problem.

■ ■ ■ ■

John seems to know exactly what I mean. He says, "It is the fact
that we burn gas the way we burn gas. That we drink coffee
and tea. That is how wealthy we are, that we can afford to eat
bananas every morning and not even think about it. Personally,
before I feel like I have a right to raise my fist in protest at anyone, I have to first improve myself."

Then he says what I have been afraid to admit to myself
ever since the WTO protests. "Life is full of hypocrisy. We are
full of hypocrisy. I think the only way to battle it is to bring
the whole revolution inside yourself. For example, I don't eat
any tropical fruit unless I am in the tropics. I don't buy
bananas. I don't drink orange juice. I feel like it is our consumption up here in the Northern Hemisphere that is causing
the rest of the world to suffer. It is our patterns, and it is our
own inabilities."

John interrupts himself with a grin, "I do drink coffee.
But, I try to drink fair-trade organic coffee that is grown in the
shade in the northern latitudes of Mexico. The bottom line is
that I am just trying to figure out how I consume."

I am impressed by John's candor. Oftentimes I resort to
feeling guilty when I buy something that is not organic or from
a chain store. But rather than wallowing in guilt, John seems to
take responsibility for his actions and learn from them. I aspire
to his awareness, and I think it is this kind of awareness that
can ultimately change how we all consume.

Backyard Homestead

There is now a "fair-trade" movement that is trying to counter the policies of "free trade" and educate consumers about how products are being produced and distributed. The Fair Trade Labeling Organizations International (FLO) is head-quartered in Germany and is the umbrella organization for Fair Trade certification initiatives in seventeen countries.[12] Fair trade works to pay a premium price directly to the producer, eliminating unnecessary middlemen, inspect labor conditions to ensure that there are no safety or child abuses, and prevent the use of agrochemicals. So John's coffee, for example, may have been grown on a cooperative farm that is run by a family instead of a large coffee plantation. Moreover, by being grown in shade versus full sun, as characterized on the plantations, the coffee plants provide a canopy of native trees, shelter for hundreds of species of birds, and prevent erosion and deforestation.

John continues, "Part of the way that I watch what I consume is through farming, because I eat mostly food that I grow myself. But I still have a long ways to go. Sometimes I feel like the best thing I could do in farming is to eventually get down to even a smaller scale, where I am only using an acre. To me, it is not how can I check out my systems to make more money, it is how I can consume less."

It is as though John is trying to rely on a backyard homestead rather than a global one. He says, "If I had just an acre I could do a lot of cover-cropping, have time to introduce animals, and have time to get all of my fertility from the land around me. I could grow grass, cut it, and compost it rather than having a truckload of chicken manure delivered to the farm. This is all a personal challenge."

He shakes his head, "I think a lot of farmers need a wake-up call. I mean, they call themselves organic, but how different

Fields That Dream

are they from a big system anywhere else? I look at some of those huge farms in California, and I feel like they are taking the term 'organic' and cheapening it. How much fossil fuel do they burn every week to make their systems run?"

John stops, leans forward in his chair and tries to explain what he is thinking in another way. "It is like this. My mom used to hear that you are supposed to feed your kids vegetables because it makes them healthy. Then I started telling her, 'Mom, you have to start buying organic vegetables to stay healthy.' Now I am going a step further and saying, 'Mom, you have to buy LOCAL organic vegetables.' She is raising her arms and saying, 'C'mon!'"

He laughs for a minute and then continues earnestly, "But that is what I believe. I believe that there are degrees of anything we do in life, and it is not enough to just eat organically, in my opinion. Local organic is the only answer. Eating your food seasonally."

■ ■ ■ ■

John didn't set out to become a farmer. A biology major in college, he was planning on a career in native habitat restoration. But during college he worked for a man who had been farming for more than twenty years, and John began to see small-scale agriculture as another way to make a living.

John shrugs his shoulders and remarks that a career in prairie restoration would probably have been more financially lucrative than growing vegetables. "I absolutely love farming, and I wouldn't exchange it for anything right now. But I am getting to this age where I am thinking about having a family, and all of a sudden the realities begin to look a little grimmer when you think about making money. But I wonder if my wife and I could raise a family on this kind of income? I know we physically could, but would we want to?"

John goes through the world with his eyes open. He honestly and carefully looks at how all of his actions affect the lives of not just humans, but life all around him. I wonder if that is the biologist in him coming out. It seems to me that his awareness is both his greatest gift and struggle.

He says thoughtfully, "I felt a lot more satisfied doing prairie restoration than I do farming. I think that farming is not intrinsically good for the environment. Sure, there are organic farms, but some of them, like I said, are huge and don't seem that different from the gigantic conventional industrial farms. You see, what we are doing by farming is taking away habitat for nature. We are causing birds to have to find their home elsewhere."

I am beginning to get John's drift. It seems as though he is taking the concept of Manifest Destiny and homesteading and putting it under a microscope. He is taking a look at all life-forms affected by consumerism, and it is painful. John says, "Farming is not necessarily an altruistic profession. The fact that it feeds people and that it keeps them from buying something that was created in a different way halfway around the world is a good thing. But just sustaining the natural ecosystem seems a far better pursuit. It feels like it is more pure."

"But," John says, frustrated, "Sometimes I feel that farming is a symbol of our domination over the land. When I am out there on a tractor I know sometimes that I am mowing birds' nests. I've got a female sparrow chirping away at me, telling me that I am about to mow down her nest. Am I going to get off every time and check the nest and then mow around it? No, I can't do that. I know that in farming, even if I try and create habitat, ultimately what happens to the field is up to my whims."

John lets out a big sigh. "Basically, I understand that the farmland that I am on is not what it was like pre–white settlement. It is all second growth. All of those big cottonwoods were probably never there. So, in that way, I am taking what I have got and trying to make it better."

He continues with an edge of hope in his voice. "What I want to do is figure out if it is possible to both farm and do native restoration. At one point I knew just about every plant species that existed in the tall grass prairie. I was really excited about that knowledge; it is just more complex than farming in some ways. It is very interrelated and very, very cool."

John realizes that he's got to get going. He is meeting a friend for dinner. I take a last swig of World Peace tea and then we head out of the teahouse together. We dodge fat raindrops as I walk him to a tattered red pickup truck where a big black dog is very happy to see him. I tell him I'll see him soon at the market for my weekly bag of greens.

After he gets in his truck and it sputters away, I think about what John said at the beginning of our talk: "The tougher you are the further you'll go." I don't think he was referring to himself when he said that, but he should have been. John is doing what many of us fear. He is teaching us how to examine the state of the world by looking at our own patterns of consumption full in the face. He is trying to make us understand that how and what we consume can be felt by small farmers, factory workers, and governments around the world. He is trying to live simply and honestly. If that isn't tough, I don't know what is.

Chapter Six:
Borders

Borders are set up to define the places that are safe and unsafe, to distinguish us from them. A border is a dividing line, a narrow strip along a steep edge. ... It is in a constant state of transition.[1]—*Gloria Anzaldúa,* Borderlands, La Frontera: The New Mestiza

I had forgotten that I was in a borderland. I am reminded that I am on the edge of the United States when I see a lone Border Patrol car poking along. He has just pulled out of the drive-through espresso stand and is blowing on his latté in an effort to cool it down. A quail waddles quickly after the car with a bundle of hay in her mouth and her curlicue headdress bobbing madly.

The air smells of evergreen trees and the thick, gray rain clouds that blanket the sky. Up toward Canada, farmlands still outstrip the malls, and the fields are lush with bright yellow dandelions and armies of green tulips standing at attention. I am headed out to see Gretchen at Alm Hill Gardens and meet her farm manager, Juana Lopez.

I find Juana in a row of sweet peas. She and her sister-in-law Carmela are picking flowers for Alm Hill Gardens to sell at the market. Gretchen introduces us in Spanish and we all stand around a little awkwardly. Or maybe it is just me that feels awkward. My high school Spanish is rusty and I am a bit nervous.

I was originally supposed to have lunch with Juana, but since it has been raining all morning, everybody has just started working. Gretchen suggests that I follow Juana and Carmela out to the field where they are going to pick tulips. We trundle off toward an ancient farm truck because the field is down the road on a different piece of property. When we pile in, I notice that there is a hole in the floor and I can watch the road whiz past.

Juana chuckles and says, "Yes, we have built-in air-conditioning." With Juana's laughter, the ice is broken and conversation begins to flow easily.

Juana's glossy black hair is pulled back into a half ponytail and she is strong, with a round, open face. Carmela is very quiet, a little lighter skinned than Juana, and has thick brown curls that tumble out from under her baseball cap. This is only her second day on the job, and I don't think she speaks any English. Juana speaks very good English, but prefers to speak in Spanish if given the option. Fortunately I know just enough Spanish to get by.

I learn that Juana has been working at Alm Hill Gardens for nine years. She explains that when she first came to the United States twelve years ago, she worked on a large conventional farm picking strawberries.

She says, "I didn't like it much because the whole time I was there the bosses didn't know me. It was only that we, the workers, need money, and they, the owners, need workers. That is it. There is no relationship at all. You don't even really see the bosses except when they drive by in their nice cars, and they never talk to the workers unless they have to."

She continues, "Here, on a small farm, the bosses are near, we work side by side. We talk to each other. Gretchen has learned Spanish and Ben knows a little. We make jokes, we have fun with each other. If I ever have a problem, Gretchen and Ben

Fields That Dream

will help me. We have a friendship. I guess that is the reason I have worked here for nine years!"

Juana laughs as she pulls off the road and we bump over a muddy field until the truck lurches forward, heaves a big sigh, and dies.

We hop out by a large bed of tulips, and Juana instructs Carmela on how to snap the stems near the root of the plant and pile them into large black crates. Carmela starts on one end of the field and Juana and I are on the other. I feel slightly ridiculous trudging after her like a shadow, balancing a crate and tape recorder in my arms.

Juana continues to tell me about working on the strawberry farm. "*Pues*, you work a lot of hours. We were paid by the amount we picked, not by the hour. On the large farms we worked and worked and worked and we still didn't make very much money. I would work for ten hours and make only fifty dollars. But, since my husband worked there, that is where I had to work."

Juana pauses to grab a new crate and jump over the row of tulips to her right. I follow her lead so we can walk side by side. I ask her about the use of pesticides on large strawberry farms.

She pauses for a moment with her hand on her hip. "I don't know much about pesticides, but I do know that some of the problems come because the bosses don't use a lot of caution, and a lot of times the workers can't communicate with the bosses because they don't speak English. There are not a lot of precautions, and because a lot of times the instructions on the pesticide containers are written in English, nobody understands them."

We continue moving down the row and Juana says, "People farm with pesticides because they are so easy. I mean, you can just kill what you don't want. But the danger is because the workers and the bosses aren't communicating about it. The workers need more instruction. I think now, since so many children have been getting sick, the bosses around here are trying to instruct the workers more than before."

■ ■ ◢ ◨

Juana's experience on a large conventional farm was not an unusual one. Although there has been some movement to increase the safety of farmworkers using pesticides on strawberry and grape farms in particular, the Pesticide Action Network reports that "Agricultural workers face greater threat of suffering from pesticide-related illnesses—including acute poisonings and long-term effects such as cancer and birth defects—than any other sector of society."[2]

Farmworkers and their families are exposed to pesticides through all aspects of agricultural labor, from preparing the fields, weeding, harvesting, and handling the product to living in or near treated fields.[3]

And pesticides are just one of the problems that many workers face. They must also contend with substandard or no housing, black market wages, sharecropping frauds, and little access to health care, education, or legal assistance. While there have been many advances in workers' rights and safety due to the efforts of the United Farm Workers of America, the rise in the number of migrant farmworkers ready to work without joining the union has grown exponentially in the last twenty years. Increased immigration and an oversurplus of laborers have driven down wages by 20 to 25 percent since the 1980s.[4] But for people fleeing the civil wars and upheaval of Central America and the extreme poverty of rural Mexico, America still holds hope for a better future.

Fields That Dream

Juana's family has long been part of the exodus from Mexico to the United States. She tells me that she first came to the United States when she was twenty-one years old, but her father started coming north when Juana was just a little girl. Juana's mother had fifteen children, but only nine survived. Only two of the children remain in Mexico. Juana explains that they both have good jobs with the government, which is one of the few ways that people prosper economically in Mexico.

But the government jobs are very limited, and Juana says, "I came here because there are many more opportunities in the United States. I came with the American Dream."

She shakes her head, brushing a stray hair out of her eyes. "You see, at times there is work in Mexico, but there is never a lot of money. When we come to America, we can earn here in one hour what someone would earn in Mexico for a whole day's work. We will earn maybe five dollars for a whole day of work in Mexico. That is only enough to buy beans and tortillas, not enough to buy shoes or clothes for the family. Beans and tortillas, nothing else. There are many more opportunities here. And when you have an opportunity, you have to take advantage of it. *Por ejemplo*, when I was young, I studied, but I stopped going to school when I was eleven or twelve. Here, all of my children can finish high school."

Juana maintains that her first seven years in the United States were very hard. "I didn't feel free here. I didn't feel comfortable. Now, yes, because I have my children and I feel more independent. When I first came here I didn't know how to drive, I didn't speak any English, and I totally depended on my husband. Now I know how to drive, so I can work where I want to. It is much better. It was very hard to learn English because I didn't have the chance to go to school. It is hard to work in the fields and then go to school. When we come home we are too tired to do anything like go to a class."

Juana, who is thirty-four, has four children ranging in age from fifteen to three, all of whom are American citizens. She herself became a citizen two years ago.

I ask her about the differences between the United States and Mexico. Juana laughs and says instantly, "The food!"

She continues, mocking the incredulity she felt upon her first introduction to the Northwest. "When we first came to Lynden, Washington, we couldn't find tortillas! We went in the store and we couldn't buy tortillas! And we couldn't find chiles! We thought, 'What are we going to eat?'"

She pauses, both of us doubling over with laughter, "And then, when we could finally find tortillas, we would buy as many as we could and freeze them because we didn't know when we would be able to buy them again!"

She stops to think seriously about the differences between the two countries and then says, "The culture. Women here in the United States, we are more independent. In Mexico, we are in the house. My sister the teacher, she is the only one of the whole town who is not in the house. Family life is different as well. In Mexico, if you are eighteen and not married, you stay in the house. Here, as soon as you are eighteen, you are out of the house. And in Mexico, even if you are forty years old, but not married, you are in the house with your parents until they die.

"Also," Juana says, "another difference between the two cultures is that when American teenagers work, the money is just for themselves. Not my son. When he works, he earns money for the family. If he needs shoes, socks, or T-shirts, he will buy

Fields That Dream

them, but then he gives the rest of the money to the family."

We pause to take another load of tulips to the truck, and Juana checks in with Carmela to see how she is doing. I look around, enjoying the crisp air, the feeble but patient sun, and the rapid-fire conversation of the birds that are speeding around the field like mini-torpedoes.

Juana returns, and we start down a new row. She tells me that her husband, Nabor, first left home when he was twelve years old. "First he looked for work in Mexico, and then he went to the border because there is more work. He picked cotton on the border for a while, and then he came to the U.S. so he could earn more money. He has been here ever since." Although Nabor has a green card and can speak English, he doesn't have citizenship. Juana explains, "He never had a lot of education when he was a kid. He hasn't ever had time to study, he has always been working, so he doesn't really know how to read and write. But he has been in the U.S. for so long that he has helped the whole family find jobs. He has sent Gretchen almost everybody that has ever worked on this farm."

Gretchen told me earlier that Juana's achievement of citizenship is unusual. She explained that people of Juana's age who come to the States as young adults rarely pursue citizenship because it is very difficult. I ask Juana how she feels about the fact that so many workers do not have American citizenship.

She replies, "It is very sad. When people come without help, without opportunity, they can't legally work because they don't have documents. The laws make it so that there is nothing that we, as citizens, can do to help, not even for family. It is terrible to see. And I know that so many people die when they are crossing the border into the United States. They come here for work and then they need to travel home to be with their families. It is very dangerous. For example, my brother, he doesn't have papers. His wife and children are in Mexico, but he can't go home because crossing the border is too dangerous. If he does go home, there is no work and no way to feed his family."

Trying to live between two countries is not a recent dilemma. Ever since the development of the West Coast agricultural industry, Japanese, Filipino, and Mexican workers have been a mainstay of labor, but they have never been welcome to call the United States home.[5] From the Chinese Exclusion Act of 1882, which suspended the immigration of Chinese workers for ten years (later extended to an indefinite period of time), to the Alien Lands Act of 1913, which prevented noncitizens from buying or owning land in California, the American government has institutionalized the paradox of relying on cheap foreign labor while denying workers the basic rights of citizenship or safe passage between two countries. Another example is the Bracero Program, which was initiated in 1942 to allay the fears of growers whose Caucasian laborers were moving into more war-related industries. For the following twenty years, between 4 and 5 million Mexican farmworkers were recruited under government sanctions to work in the fields, but only as temporary workers.[6]

Currently there are 1.5 million seasonal farmworkers in the United States, 700,000 of whom are migrants and a majority of whom are undocumented illegal immigrants.[7] These workers, who provide a seemingly endless supply of cheap labor, are the backbone of the American agricultural economy, yet they live in constant fear of being apprehended by immigration authorities and are subjected to intense discrimination, some of it through legislation. The most recent legislation against immigrants was California's Proposition 187, "which intended to deny public

Fields That Dream

school education and health care to undocumented immigrants and their children."[8] The legislation was the culmination of typical anti-immigrant sentiment that illegal "aliens" are stealing jobs from American citizens and taking advantage of government resources such as welfare and subsidized health care.[9]

Juana reiterates, her voice lowering with frustration, "America depends on the Mexican workers but we don't have any rights. Workers have to be able to get papers because it is getting more and more dangerous at the border."

Crossing the border can be a harrowing journey. A couple of weeks ago I was visiting a friend in Arizona who lives just a few miles north of the border, a toothed line of barbed wire. There was a blockade on the highway: sirens, the crackle of radios, and the serious air of a crisis. Anyone who was brown was stopped. As my friend and I were waved through with a smile and a nod, she told me that this sort of thing happens all the time. It was clear who the "aliens" are.

The border is virtually a war zone. The United States Immigration and Naturalization Services (INS) uses "helicopters with searchlights, infrared sensors hidden in bushes and motion detectors buried in the desert" to stem the tide of immigration.[10] The INS has 26,000 employees, nearly half of whom are posted along the United States/Mexico border, and an $800 million budget.[11] And yet it is like trying to hold back the ocean—an estimated 1.5 to 2.5 million people enter the United States illegally every year.[12]

■ ■ ■ ■

Carmela has finished picking and starts gathering up the crates and various tools, so Juana and I end our conversation. The wind has picked up and we load the rest of the tulips into the truck and then into the storage shed near the road.

On the ride back to the main farm, Juana tells me about her dreams for the future. "*Bueno*," she says, smiling. "I wish that my children have good opportunities, that they have good principles, and when they are ready to leave home, when they are ready to make their own life, I would like to retire in Mexico. Nabor and I have already started to build a house there near our family. We will have a good retirement. Work here, rest there."

Gretchen meets us back at the main farm. I take a few pictures and thank Juana for her time. Gretchen and Juana are already talking about the next task at hand, and so I leave without any long good-byes.

As I drive back out to the main road, I admit to myself, now that the interview is over, that it was something I had wanted to avoid. The truth is, I felt silent embarrassment and shame that virtually all of the food, conventionally and organically grown, on America's table has been put there by brown hands and that the same people who feed America are under siege while they are here. But I don't understand why this feeling made me create a border in my mind—one that helped me ignore the fact that injustice is part of every meal we eat.

Most of the Mexican farmworking population is unwelcome, poor, and invisible, without health care and social security. However, many illegal immigrants show fake IDs and social security cards to employers. This enables immigrants to obtain jobs more easily, and employers the ability to claim that they believed all of their workers to be legal.[13] Consequently, these undocumented workers have social security and Medicare taken out of their paychecks—services that they, as noncitizens, will never have access to. In fact, it is estimated

Fields That Dream

that the approximately 7 million illegal immigrants in the
United States are providing the social security system with a
subsidy of almost $7 billion a year.[14]

But Americans are angry that Mexicans are flooding the
country, and yet without them, the cost of food would rise
exponentially. A union organizer for Mexican farmworkers
says, "Are we going to put all the white people out there with
a bucket on their hip? I don't think so."[15]

Alm Hill Gardens and other small farms across the
country are setting the standard for a humane and dignified
approach to labor. Juana is treated with respect, good pay, safe
housing, and tremendous opportunity for her children.

But for tens of thousands of other workers, there is still
such a long, long way to go. Will there ever come a time when
the power scales will balance? Will the borders of skin color
ever be erased? Will farmworkers ever be revered for feeding
a nation? I head south on the freeway back toward Seattle,
Canada's "Peace Arch," with stretches of manicured lawn and
seasonal flower beds, behind me, and Mexico, with a ragged
border of questions, ahead.

Chapter Seven: Mountain People

"Fahma," said Mai Chia, pointing to herself. "Fahmah." She led me by the arm over to the wall and tapped it meaningfully. "Fahmah," she said again. She was, she seemed to be saying, a farmer surrounded by concrete.[1] – Nancy D. Donnelly, Changing Lives of Refugee Hmong Women

Joua Pao Yang speaks quiet English. "When the Hmong people came to this country, we were not used to cities. Most of us had never been to school, never knew about factory jobs or working at a company.

There were no companies up the mountain. We worked the farm. We planted corn and rice. We didn't know how to live in America; we were not familiar with this culture. The Hmong, we are mountain people."

Joua Pao pauses and rests his elbows on the arms of the enormous brown easy chair where he sits. His dark, handsome face is kind and open, and lined with experiences I will never hear. I look around. We are not in the mountains now. Oceans of wet, gray concrete surround us. It is a very dark evening in the middle of January, and I am sitting on a wide couch in the Yang family home in a small suburb about fifteen minutes north of Seattle.

It feels strange to be sitting in their living room in the middle of winter. I usually see the members of the Yang family nearly buried in a brilliant avalanche of flowers at the farmers' market. Sua Yang Farm is one of a network of more than sixty Hmong-operated farms, run mostly by the women of the families. They specialize in growing flowers, and the local farmers' markets are cloaked in lush tapestries of their lilies, delphinium, dahlias, and sunflowers. Sua Yang and the other farmers have amazed Seattle with bouquets of color and form woven together in a way that is reminiscent of the rich textile work for which the Hmong are known.

The Hmong, who began farming in this area soon after their arrival in America during the early 1980s, have richly diversified the ethnicity of local family farmers. Before World War II, the Seattle area was teeming with Japanese, Filipino, and Italian American farmers. In recent years, however, Caucasian farmers have become the overwhelming majority. Now, the Hmong and the few other Southeast Asian farmers in the area bring to the Seattle markets a much-needed abundance of flowers and a wide array of Asian greens new to the American palate.

Most of the Hmong farmers congregate around the Snoqualmie River Valley near the towns of Carnation and Duvall. A generation ago, this valley was filled with small dairy farms, and now much of the agricultural land has been developed into suburbs or gone fallow. Mark Musick, farmer liaison for Seattle's Pike Place Market, says, "They [the Hmong] have played a major role in revitalizing the market and agriculture in eastern

Fields That Dream

King County. These people are returning agriculture to places where it has been abandoned."[2]

When I arrive at the Yangs' house, an elderly woman, stooped with age and wearing a pink cardigan over a flowered housedress, opens the door. She smiles warmly, takes my coat, and leads me into the living room. I slip off my shoes, leave them near the neat stack of others by the door, and take a seat in the living room. Joua Pao enters soon after, greeting me jovially. He introduces me to his mother, who then heads back into the kitchen.

Joua Pao and I are now deep in conversation. He is being very patient with me as he helps me navigate the story of how he and his family first came to this country. His words wrap around each other like the delicate threads of the traditional Hmong story cloths I have seen only in books. While we talk, his wife, Sua, enters the room and sits perched on a nearby love seat, silent. She is a beautiful woman with high, rosy cheeks, weather-swept skin, and thick black hair that she keeps loosely tied at the nape of her neck. I try to draw her into the conversation and soon realize that, in addition to Sua not speaking much English, it is probably customary for men to do the talking.

■ ■ ■ ■

From what I can tell, the Hmong people have always been on the run. Originally they were from China, but because the Hmong were being "ethnically cleansed" by the ruling Chinese as early as 3,000 B.C., they eventually fled to the mountainous regions of Vietnam, Burma, Thailand, and Laos. Arriving in Laos wasn't much of a homecoming; they were despised as a racial minority and "for the century and a half that the Hmong

Mountain People

lived in Laos, they were essentially outsiders, tolerated as long as they were largely unseen."[3] This isolation did not seem to bother the Hmong, who prospered in the Laotian mountains, continuing their agricultural way of life until Cold War politics of the twentieth century stormed into their reality like an unrelenting blizzard.

As Southeast Asia struggled to come out from under colonial rule, its countries fell under the powerful influences of the United States and the Soviet Union. Korea, Vietnam, and Laos were divided between north and south, and vicious civil wars were fueled by either American or Soviet ammunition. While America's involvement in Vietnam was overt, there was a brutal "secret war" happening in Laos, engineered by the CIA. Seen as strategic land, Laos was overrun by the warring Vietnamese, who blazed trails through the mountains and Hmong homelands. The CIA began recruiting and training Hmong men as guerrilla fighters to gather information about the North Vietnamese as well as fighting the Laotian communist regime.

The Hmong were not particularly on one side or the other. Lillian Faderman, who collected the oral histories of Hmong in America, writes:

> Most Hmong knew very little about ideology or differences between Communism and Capitalism; the communal farming with which they often happily engaged can even be considered a kind of "communism." But they were convinced by those that recruited them that the North Vietnamese would invade them, take their homes, and subjugate them if they did not fight.[4]

Fields That Dream

America also made what turned out to be empty promises of food, medical supplies, and other assistance if the Hmong would fight against communism. However, this secret war left the Hmong with a casualty rate that was "proportionally ten times higher than American losses in Vietnam," and by 1971, Hmong families were bereft of men and boys down to the age of ten.[5]

Once the communists took over Laos, the Hmong were persecuted as traitors, hunted and killed or taken to "re-education" camps where many often died under poor conditions. Thousands of Hmong fled to the jungles in a desperate attempt to escape to Thailand, and many more died crossing the treacherous Mekong River that divides Thailand from Laos. The persecution of the Hmong in Laos continues to this day. The Communist Lao People's Democratic Republic (LPDR), which receives American economic and foreign assistance, is still hunting down ethnic Hmong and the former Hmong guerilla fighters that were trained by the CIA. The Lao Human Rights Council reports that the LPDR has killed more than 300,000 people in Laos since 1975.[6]

The Hmong are not the only indigenous farming culture that was destroyed under the vice grip of the Cold War. United States' foreign policy, some of it under the guise of agricultural innovation, has ripped millions of people from their farming legacies and created a whole generation of refugees. The wars that have torn apart Vietnam, Cambodia, Laos, Ethiopia, and Eritrea are rooted in centuries of colonialism, invasion, and power struggles between communist and capitalist countries—struggles that have resulted in poverty and hunger.

This world hunger prompted scientists in the United States and Europe to mastermind the "Green Revolution" during the

1940s through the 1960s. The idea of the Green Revolution was to increase food supplies as quickly as possible by developing high-yielding varieties of staple crops such as corn, wheat, and cotton. These new crops were designed to replace the more traditional strains of seed that were being grown worldwide. In many cases, however, the new seeds brought to bear a different set of problems. Forerunners of genetically modified organisms (GMOs), they responded best to synthetic nitrogen-based fertilizers, which were not commonly used and had to be bought, most often, from foreign suppliers. Furthermore, the suggested practice of monoculture, the planting of one crop instead of many, encouraged more pests because the pests had more of the same crop to prey upon. Eventually, only the financially prosperous farmers who had money to invest in the new seeds, fertilizers, and pesticides succeeded, and the poorer farmers were worse off than when they started.

Some scholars argue that while the Green Revolution had obvious humanitarian goals, it was really a response to Cold War politics in which "hunger was the ally of the communists," and that "a major part of the impetus behind promotion of the green revolution lay in the desire to forestall revolutions of another color."[7] Indeed, all across the Southern Hemisphere there were rumblings of unrest that eventually exploded into various uprisings, civil wars, and revolutions.

The Green Revolution never solved the problems of world hunger, poverty, or social unrest since it focused only on increased crop production and foreign technology. It didn't focus on rebuilding the environment after the ravages of war or examining corrupt political infrastructures that result in unfair distribution of land and water. Neither did it focus on social justice issues such as educational and health disparities that are

Fields That Dream

often at the root of poverty and violence. Instead, the Green Revolution continued to dislodge traditional farming cultures that had been feeding families for thousands of years and opened the world market to agrochemical corporations. These petroleum-based companies, such as Shell Oil and Chevron, still prosper because the new strains of seeds are dependent on fertilizers, petrochemical-based pesticides, and farm machinery.[8]

■ ■ ■ ■

Communities of Hmong and other refugees across the United States and Europe are working to recultivate an agricultural heritage and create bonds with their communities at large, fostering a self-sufficiency that was destroyed by the trauma of war. One example of these efforts is the Indochinese Farmers' Association in Seattle, which was started in 1983.

Joua Pao says, "Since almost all of the Hmong people were farmers, they had a hard time getting factory jobs because they had no experience. A group of refugees and some people from Seattle got together and wrote a grant to the state for equipment, and the county gave them eighteen acres of land in Woodinville." It was through Joua Pao's connection with the Indochinese Farmer's Association that Sua Yang Farm was born in 1987.

But it has been a treacherous journey for the Yangs. After five years of barely surviving in a refugee camp, in 1980 Joua Pao, Sua, and their six children left for Seattle. They were among the 300,000 Hmong who have resettled in the United States in the last twenty-five years. He says, "When our family came to the United States, we were sponsored by a Presbyterian church. We lived with an American family for a couple of weeks, and we had several other people help us survive. People came to our house to teach us English and about shopping, cleaning, and cooking on electric stoves, so we had a chance to learn something before we started out on our own."

When they first arrived, Joua Pao was able to attend school full time for a year while Sua stayed home with the children. For the following two years, he balanced working full time and carrying a full load of classes. In 1985 the archbishop asked Joua Pao to work with other Laotian refugees at church as a community organizer and advocate, and he has been there ever since.

Many Hmong have left the area for different parts of the country, and the once vibrant association now only has three families farming on the original eighteen acres and an additional twenty-five acres that they have leased just north of the county property. Hmong populations are now concentrated in Minnesota and Wisconsin. Because the Hmong are a tribal people, many migrated from California and Washington to be with their community leaders. Others left because of welfare reforms and high rates of unemployment.

Joua Pao explains more about the farm. "My wife works full time at the farm, and a couple of our daughters help as well. Since I have my other job, I can only help on weekends. Sometimes I sell at the farmers' market, and I help with the plowing and tilling of the land."

Joua Pao says that Sua and the others didn't start out growing flowers. "We never grew flowers in Laos." He laughs, "I don't think Asian countries care much about flowers! When we first came to this country we started growing produce, but at the markets we saw people go and buy flowers, and we saw how many flower shops there were around. We wondered, 'Why do so many people eat flowers and not vegetables?'"

Joua Pao erupts in a peal of laughter. Sua follows suit, apparently understanding much more of the conversation than

I had thought. Grandma peeks her head around the corner of the kitchen to see what all the commotion is about and comes and sits next to her daughter-in-law. Sua translates for her and she too breaks into a rush of giggles.

The front door opens and a teenage girl dashes in from the wet night, shaking the water off her coat and her long black hair. She takes her shoes off, and Joua Pao introduces me to his youngest daughter. She stands, polite but taut, in the living room, ready to spring like a fox into the safety of her bedroom. "Nice to meet you," she says. Moments after she retreats we hear loud country music seeping from the crack under the door.

Joua Pao remarks, "This farming is very good for the children."

I must look puzzled, and so he continues. "It was helpful for us economically to start the farm, but more importantly, farming is one of the things that keeps the family tied to tradition."

"You see," Joua Pao leans forward in his seat, "I work with people in the community right now where many of the families are not farming, and their children did not graduate from high school. In these families, the children are out of the house with friends more than they are at home with the family. So many kids join gangs or just don't go to school. Some find a job at McDonald's and that is it. These kids don't think about their future at all, and that is very tough for the families that have four, five, six, or seven children. But the farm keeps a family together. When there is a farm, the children work with the parents, they learn about Hmong culture and our tradition of farming."

I have heard about such strife in Hmong families. Although given refugee status, many Hmong arrived in this country without sponsors and were handed an apartment but minimal instruction. These were a self-sufficient tribal people who had been ripped from land they based their entire existence on, and they were plopped into a country of supermarkets, electric stoves, and concrete. After suffering such intense trauma in Laos and then in the camps, America was the ultimate shock to the system. Depression and post-traumatic stress disorder still run rampant, and 80 percent of Hmong immigrants live below the poverty line.[9] For many of the older generation, education has been inaccessible. From 1972 until 2001, only 126 Hmong people among hundreds of thousands received doctoral or other professional degrees. However, the numbers of young people in school are growing. In 2001 there were approximately 6,500 Hmong American students in college.[10]

Joua Pao gesticulates with both hands for emphasis, "When we were a farming people we lived a much happier life. During my life in Laos, I never saw any divorce. I remember my father and grandfather saying that in our Hmong culture we don't have divorce, maybe for every thousand there is one divorce. Not like in this country. It is so hard to live between two cultures, and everything has changed so much."

He continues, "For this generation of children who were born in Laos and who grew up here, I don't think they will forget or reject being Hmong. But I think that in two or three generations down the road our culture may disappear. Because of technology and all of the new jobs, I don't think that generation will be interested in farming. It is so sad. In the future, when all the parents and grandparents are gone, then they will forget Hmong culture."

Joua Pao speaks sadly but resolutely, as if he has seen the future and knows that it is inevitable. I don't think he has anything else he wants to say. I thank Joua Pao and Sua for their hospitality and we walk toward the door. Joua Pao suggests that I talk to his daughter Xee. "She will have lots to talk to you about," he says, and Sua nods in agreement. Grandma comes and gives me a hug good-bye, patting my back as her small arms wrap around me.

It is two weeks later and I am sitting in Xee Yang Schell's living room. She lives with her husband, Joe, in a large house that still has a shiny new feeling. The living room is cozy and bright and we are sipping cups of tea. Xee, who is as beautiful as her mother, sits on the couch next to me with her legs tucked under her graceful frame. I have met Xee before at the University District Market, and from the moment I laid eyes on her I felt like we could be friends.

Xee is telling me about when her parents started the farm. "Oh yeah," says Xee, rolling her eyes, "when my mother started farming we all hated it. She started the farm when I was around fifteen. All of us kids helped her during the summertime. I mean, there was never any question that we wouldn't help. Farming does allow Hmong people to keep most of their families together because the family, whether they want to or not, will help each other out."

Xee recounts that after high school she went to Seattle University where she studied accounting. She took the Certified Public Accountant exam and went to work in an office for a while.

She shrugs, "I guess it was just never really me. Office work was such a new world, so completely different from farming! It felt so rigid. I was cooped up inside all day looking at computers, and after a while my head just started spinning. I started feeling like I was being caged, and I wanted to be outside where all of the life was. I think farmers in general like their independence."

While Xee took time off to figure out what she wanted to do, she started helping her mother with the farm. Xee laughs, "I guess I got attached to the farm, and my mother needed me, so I just decided to stay. And here I am!"

I can see that Xee is a powerhouse who almost hums with energy even while sitting quietly on the couch. She outlines her duties on the farm: "I order all the seeds, keep the records, and do a lot of the selling and arranging of bouquets, and I do a little of the harvesting."

"I am not a very good farmer," Xee admits, placing her mug on the coffee table in front of us. "But I am a good seller. I am good with people. My mother, on the other hand, doesn't like dealing with customers. Her English is very limited, and sometimes I think she feels intimidated. My mother doesn't read or write very well, but she is one of the best farmers that you will ever meet. It is her territory. It is her area of expertise."

Xee thinks for a moment, shifting her legs out from under her. "I think that farming is really good for the older generation. It keeps their practices alive and it allows them to continue doing something that they love. They love it because it is such a part of them. Being able to farm here is like being back in their peaceful villages in Laos. I think maybe they can kind of escape to a world that they know."

I smile to myself, realizing that Xee and her father think very similarly. Joua Pao, however, thinks of farming as a way to preserve a culture for the younger generation, and Xee seems to think that it is preserving a way of life for the older. Perhaps farming is serving both purposes.

100 Fields That Dream

Xee continues, thinking now about all of the Hmong women who congregate to sell at the various farmers' markets around town. "It is very good for all of the ladies to see each other so regularly. It is like when they lived in Laos, when they all lived in the same villages and would see each other every day. If they had separate jobs at factories or offices, they would never see each other, and I think that could be very isolating for them."

At the farmers' markets there is a wonderful mix of generations. Often I see the children of the farmers selling flowers and produce while the women are in the back, fingers flying as they make bouquets. Xee tells me that the women, who are the primary farmers, range in age from thirty to sixty. She explains that even though some women may just be a few years older than she, there is a big difference between them.

"My older sister is a good example," observes Xee. "She was a teenager when we came to the United States. She didn't finish high school, had a much harder time learning English, and she got married young, like many Hmong girls do. Her outlook on life is definitely more like someone from Laos, where I look at life more as an American. My sister decided that she would help my mother with the farming. I don't know if she likes it or not. Her husband works full time, they have children, and her in-laws live with them, so they help take care of the children sometimes. I think it is hard for her, but farming is something she can do that is flexible."

Our conversation pauses for a moment, and I am struck by how many experiences can be had in just one family. It seems that Xee is a bridge between the two generations. Her older siblings fall on the side of her parents, while her youngest sister, who is just sixteen, is growing up in an entirely different world.

It must be hard to navigate through so many distinct realities in one family.

Xee expresses some of the same concerns as her father: "People of my generation don't enjoy working with flowers and they don't enjoy being out on the farm. Most of the people my age are out there working for Boeing or Microsoft, or they are doing cashiering jobs or working in restaurants. I don't think they see the value of the farm or working with their families. It is so much hard work, and if you don't love it, you can't do it."

I mention to Xee what her father said about the hardships in so many modern Hmong families. She agrees. "I think many Hmong parents feel very sad, like 'what kind of a curse is this to be thrown into this country where my children don't listen to me and they don't care about our culture or where they come from?' It is very sad for the older Hmong, and I think the younger people feel it too, but they don't know how to express it. They can't communicate with their parents. It is a terrible thing, whether they recognize it or not. There is so much that is a part of them that they will lose by not being connected to their roots and to their parents and to their past. You can't completely break away from your roots. You can't just forge into the future without knowing your past."

Xee tells me that she is trying to arrange a trip with her parents to go back to Thailand and Laos. "I want to hear the stories of how we lived and what happened there. I want to hike the trails they hiked and for them to be able to tell me how it used to be when they were growing up."

Xee's husband emerges from the basement. He is a tall, handsome Caucasian man, and he greets me warmly. It is getting late and all of us need to get up early for work. I take a last gulp of my tea and put my jacket on.

Fields That Dream

I am very sad as I drive home, and I find that the sadness lingers, even at the crest of summer as I buy an armful of brilliant zinnias from Xee and Sua at the farmers' market. I am upset that these bushels of flowers might be a swan song for a generation and a people.

Joua Pao's words echo in my mind, "Everything has a meaning. Everything has value. Many people do not really care about the meaning or value of what they do every day. But this farming is valuable for the Hmong. It holds great meaning. It is our background."

I remember asking Joua Pao, almost desperately, if Hmong people ever think about returning to Laos to resume their way of life. I am momentarily forgetting that the landscape and the way of life in almost all of Laos and Vietnam have been irrevocably changed by war. He says quietly, "Even if you go back or continue going forward, it will never be the same."

Xee echoes her father's words. "The Hmong are a very strong people to have gone through so much and then come to this country and just continue on. I think a lot of Hmong people still have a free spirit. Independence. I mean, I want to work on my own terms, even if I have to work twice as hard. I love the farming. It is in our blood. Our ancestors did it for millions of years. Maybe we are the last ones."

I am left with questions. Can a people rebuild a traditional way of life after the suffering of war? How can indigenous cultures survive in the face of First World technology and ambition? More specifically, can the Hmong culture be sustained and renewed on new land? Will the Hmong children of this young generation feel the call to farm? Will they remember the mountain?

Chapter Eight: Snow

Geese

When I talked to Joanie McIntyre on the phone about coming to visit Rent's Due Ranch, she told me that the farm was on a tiny island east of Camano Island. Looking on the map, I find that Camano is the largest in a small archipelago about an hour's drive north and a little bit west of Seattle. Apparently you can drive to Camano via a series of bridges that pass over this collection of islands, which in my opinion, now that I am here, looks like a bunch of land separated by large puddles or sloughs.

Joanie told me on the phone that the island where she lives used to be surrounded by a band of water 100 feet wide, but that the

county rerouted it, so now there is just a little stream. I know that in the grand scheme of technology, rerouting a mere 100 feet of water is paltry. Still, I am impressed by the change it has rendered on the land, making island status almost an illusion.

As I head out to the ranch, I notice that instead of passing dairy farms I am passing

105

Dairy Queens. Farther from the highway, there are one or two abandoned-looking farms scrunched between an ever-encroaching sea of tract homes. When I make a left onto the road that will take me to the farm, a conglomeration of gas stations and mini-marts clutters the landscape. But after about a mile or so, suburbia begins to recede. I cross the bridge delineating island from island. This tip of earth that pokes its head inches above water has far fewer homes on it and appears to be mostly fields and marshland.

Joanie McIntyre and Michael Shriver started farming organically in their backyard twenty years ago. Now they are certified organic and farm thirty-two of their forty-four acres, with sixteen acres in vegetable production and the rest devoted to berries and greenhouses.

I am greeted at Rent's Due Ranch by a huge sheepdog who starts to bark ferociously as I try to get out of the car. I roll the window down and start talking to him.

"Hey, it's okay, I am just here to see Joanie and Michael." After a couple of moments, the dog stands up with an audible sigh and backs away. I venture out of the car and he serenely leads me to the barn a few yards away where the racket of a power saw is deafening. I understand why my siege went unnoticed.

The operator of the power saw is a tall man with brownish blond hair that falls to the middle of his back. He has mischievous blue eyes, and I am able to discern a smile by the way his beard begins to creep up around the edges.

"You must be Jenny," he says in a mellow, almost languid voice. "I see that Rufus escorted you in."

Michael tells me we will find Joanie in the potting shed. As we walk, I comment that the farm is a lot bigger than I

Fields That Dream

expected. Michael explains that actually they don't have a huge amount of land, but what they do have is extremely productive. He says, "Last year we cropped four times. As opposed to conventional agriculture, which plants once a year, we are continually planting the same ground, rotating between crops we sell and cover crops like fava beans and vetch, which replenish the soil. A lot of times, when we harvest a crop, we are planting something else in that ground the same day. On average, we are growing 25,000 heads of lettuce a week."

I must look shocked by the magnitude of food that Rent's Due produces, because Joanie laughs when she looks up from her work in the potting shed and sees the stunned expression on my face.

Joanie is an absolutely lovely woman in her early forties. She is slender, extremely strong, and has long, wavy brown hair that she wears tied back in a ponytail at the nape of her neck. She has lightly tanned skin even though it is the very end of winter, sparkling deep-blue eyes, and a number of silver earrings in each ear.

She greets me and then jumps into the conversation. "When we were living in Snohomish, the garden was pretty much just for us to have fresh food."

She gestures for me to take a seat somewhere, so I hop up on one of the tables in the potting shed. It is a cold, gray day in the middle of March, and we are all bundled up. Our conversation is turning to steam as soon as the words hit the air.

"When we first started out," Joanie continues on her previous track, "we sold our extra produce at the local farmers' markets. Every year we just started doubling the size of our garden pretty consistently." Joanie shrugs, "And then we ended up here."

Snow Geese

Before they launched into farming full time, Joanie was a preschool teacher and Michael was a jack-of-all-trades. He chortles, "I have never technically had a job, I just worked every day! My mom always used to ask me when I was going to get a 'real' job. In fact, she still does!"

Joanie explains that neither of their parents ever thought they would go into farming.

"Basically," declares Michael as he walks over to a huge pile of potting soil in the corner of the shed, "we were both raised in the city without any exposure to agriculture at all. Genetically we must have had some roots in it."

Michael grabs a shovel and starts filling seedling trays with the dark, spongy soil. He lines the trays up on a nearby table and Joanie walks over and takes a stack. She rejoins me and pulls along a flat of zinnia seedlings that she gently lifts out of the tray with one hand, catching the delicate root systems with the other. She begins to transfer the separated plants into the seedling trays, which, she explains, will be tended in the greenhouse until they are ready for sale. Just in the last couple of years, Joanie and Michael have started a nursery business in order to get an earlier start to the growing season. They sell their plant starts during the first days of the farmers' markets and to a number of grocery stores throughout the season.

I ask if I can help with the planting, and Joanie hands me a wooden stake that helps separate the intricate root highways. I inquire, "Do you ever look back and feel totally surprised at where you have ended up?"

Joanie answers immediately, "I am. But Michael is really directed. I guess I always figured we would do it, but I never thought we would be this big, or in the nursery business."

"Oh, I knew it," says Michael with a sarcastic authority. "I knew everything. I knew it all from the time that I was about one year old."

Joanie smiles and I rephrase my question. "Have you been overwhelmed by the way the farm has grown?"

Michael says in a moment of seriousness, "We were constantly overwhelmed."

"But you kept doing it?" I persist.

"Well, I guess I love the torture!" Michael chortles.

Joanie smiles at her husband and then expands, "I think farming is a love/hate relationship, and love always works out. If it didn't, I would never keep going. Farming has its frustrating moments, especially when it comes to the business aspect of it. You have to be a businessperson to survive."

Michael says soberly, "Oh yeah, you have to be able to look at the books and play all sorts of financial games. There aren't a lot of economic rewards either. We have progressively made more money, but we are still living at the standards that we started with twenty years ago. Costs are so high, and we continually have to reinvest in the business. In fact, we probably had more pocket change back then."

Joanie interjects, "We didn't have four kids back then either!"

Michael maintains, "The financial rewards haven't become spectacular. You have to live at a certain standard, as opposed to different facets of society where the rewards are a lot higher for less effort."

I ask what would happen if they reinvested less, and Joanie explains that they wouldn't be able to stay competitive.

Michael elaborates, "Well, the big boys have gotten into it. The largest production farm in California is now organic, and California sets the prices. We keep making the farm larger so we can keep up, and now we are realizing that the bigger we get, the higher the stakes are."

"Now that we are getting a lot bigger, it gets really scary at times," declares Joanie. "I mean, we still don't make enough money to have health insurance or feel really comfortable about sending our kids to college."

Joanie tells me that their children range in age from ten to nineteen. The eldest is in college in eastern Washington. She argues that even though there are financial risks to farming, the kids are one of the reasons that she and Michael keep pursuing it.

Joanie elaborates, "I think the farm is good for the kids. They have a great work ethic. They have to work with each other even though sometimes they don't get along."

"And we have been home the whole time that they have been growing up," Michael chimes in. "They really help with the farm in every aspect. Sometimes it is hard to tell, because they all goof around like any other kids, but they have gotten a whole different kind of education. They have learned independence, decision making, and physical skills that they wouldn't get anywhere else."

"And," Joanie adds, "I have to hand it to them, they know that they are going to have to go to work to make a living. I think a lot of other kids growing up these days don't realize that."

Although Joanie and Michael have raised their kids with a sense of independence, Rent's Due Ranch fights for its independence every day. Joanie and Michael have spent years trying to convince banks that they are a legitimate operation.

Michael says, "The banks would barely acknowledge us because we are a farm."

Fields That Dream

Joanie says wryly, "We have always had to be creative with our financing and bookkeeping. We don't fit in anywhere. When we first started Rent's Due, the bank laughed and called us a hobby farm."

"They are sure recognizing us now though, because agriculture is disappearing around these parts," Michael says as he pierces the pile of soil with his shovel.

Joanie and Michael relate the story of losing their last farm, which they had leased, to a major suburban development. Their old farm, which had a house and barn from the turn of the century, was transformed into 136 homes that are nine feet, six inches apart.

I ask if any of the locals opposed the construction, and Michael replies, "Yeah, but it doesn't really matter because you can't really stop it anyway. There is a lot of agricultural land around here, but the next generation of kids that own it are not all that interested in farming. Then these big developers come along and offer them tons of money for the land, and that is why we are seeing farms disappearing around here."

Joanie adds, "And the growth is spreading right up to the existing farms, and then the county puts all these new regulations on the farms because, although people like the idea of living in the country, they don't want the activity of the country right next door. It is ridiculous."

Michael stands his shovel against the side of the shed and goes to check on how two of the boys are doing with their Saturday chores. Joanie decides to take me on a tour of the house. As we crunch along the gravel driveway, I ask her what it is like to have her husband as her business partner.

She says, "It is fun, and, well, it is hard. I don't think husband and wife teams are as normal in this day and age. We are with each other all day, every day. We are in each other's faces all the time, and it has its challenges, because you don't really have any privacy at all. I don't think it is necessarily a bad thing, just unusual."

After she shows me around the house, we sit at the kitchen table for a minute. I tell Joanie it is amazing that they can fit six people in what appears to be a traditional two-bedroom farmhouse. Bedrooms have been created in a walled-up porch area, and evidence of children is everywhere.

Joanie shrugs with a grin and continues our conversation. "It's funny, but even though Michael and I have a very unconventional lifestyle, I guess we still have a pretty traditional marriage. Not only do I do all the housework and cooking, but I am out here on the farm too. I do everything. I mean, I was out there in the '70s listening to what Gloria Steinem had to say and all that. But then kids came into the picture and that made a big difference."

She chuckles wryly, "Yeah, I think we women end up burning ourselves out. I reached a point in my life where I went, 'What the hell am I doing?' And then everybody had to stop and pay a little more attention because they realized that I was going to stop doing all of the cooking and cleaning. My family had to learn a lot, like doing their own laundry, because I am not doing it anymore! I can't do it all!"

Joanie pauses, smooths a few waves of hair that have fallen forward into her eyes, and pushes her chair away from the table. I follow her move and we head back outside to find Michael. As we walk, I think about how, for better or worse, we can make radical decisions in our careers and lifestyles while still maintaining very traditional roles within family systems. Joanie stops and looks at me, "You know, money doesn't mean that much to me. It hasn't been my life's ambition to get rich. Honestly, I can say that in high school all I really wanted to do was change the world! I liked politics. I was always out protesting something or other, got arrested, and those sorts of things. Farming is my way of doing the best I can to make a political change. Without a doubt. And I've always known that work done on behalf of social change won't make you rich. Neither of us ever wanted a mainstream lifestyle anyway."

■ ■ ■ ■

I don't know if twenty years ago Joanie or Michael could have ever foreseen how trying to maintain the tradition of family farming would become such a radical political action in the face of industrialized agriculture. The values of independence and decision making that they so treasure for their children are quickly coming under fire by the corporatization of agriculture, particularly the movement toward genetically modified foods (GMOs). Even the diverse crops that Rent's Due grows— heirloom varieties of flowers, herbs, and vegetables—are under threat of homogenization.

Heirloom varieties are seeds that have been passed down through families or seed companies for at least fifty years. Their qualities, such as height, color, rate of growth, and size of fruit, have been culled by home gardeners and farmers alike through crossbreeding and hybridization. What results is a rich genetic diversity of tens of thousands of crops from purple carrots to green "zebra"-striped tomatoes.

More variations of fruits and vegetables exist than most people can conceive of. Agricultural scientists Cary Fowler and Pat Mooney report that "In the United States alone, more than seven thousand named varieties of apples and over twenty-five hundred kinds of pears are known to have been grown in the last century."[1] Although serious, the loss of most of this diversity and the current reign of Red Delicious apples and Bartlett pears holds less concern than the lack of genetic diversity of the world's staple crops: wheat, corn, potatoes, and rice, which sustain 60 percent of the human population.[2]

The Green Revolution of the 1960s was one of the first steps in the erosion of the world's seed supply, as farmers across Asia, Latin America, and Africa began replacing traditional varieties of rice, wheat, and corn with the new high-yielding varieties. This switch resulted in the necessary adoption of synthetic fertilizers and pesticides on which the new seeds depended. Most of the traditional varieties of seeds had been adapted to local pests and growing conditions, which included usual amounts of rainfall, typical soil fertility, and traditional fertilizers such as manure. The introduction of pesticides and fertilizers exhausted the soil, and while yields did increase, the toll on the environment was great. Eventually farmers wanted to return to their traditional crops, but by that time the old varieties had disappeared, since very few people had utilized seed banks or managed to save the old seeds.[3] The cultivation of only a few strains of staple crops not only facilitated the loss of thousands of traditional seed varieties, it put both biodiversity of the seed supply and farming communities at risk of a catastrophic loss.

The Irish Potato Famine, a famous example of crop failure, predated the Green Revolution by 120 years; however, it is still the most potent example of the danger of growing genetically similar staple crops. The blight on potatoes continued for five years, and more than 1 million people died of starvation.[4] More-recent crop failures have stricken the United States.

Fields That Dream

During the 1940s, 80 percent of the oat crop was lost due to a single pest, and a corn blight killed a large percentage of the nation's corn crop in the 1970s.[5]

Despite concern over limiting seed diversity, the legacy of the Green Revolution continues with the rapid development of biotechnology. Biotechnology companies continue to tout genetic modification of plants (and now animals) as a way to end world hunger, lessen the use of pesticides and herbicides, and provide farmers with the promise of a successful harvest. These new methods have been dubbed the Green Revolution II.

Biotechnology corporations and the U.S. Department of Agriculture suggest that genetic engineering is on the same continuum as plant breeding techniques that have been used for centuries. However, upon closer examination, this new technology seems more of a departure from conventional plant breeding than an extension of it. Crossbreeding means refining characteristics of plants by using traits from other plants in the same species or genus, while genetic engineering implants seeds with genetic information from unrelated species to create plants with unique and specific agronomic traits. For example, some seeds are engineered with pesticides already in them, so that just eating a leaf of the plant would kill the targeted insect. Other seeds are integrated with genes that enhance shelf life or toughen the skin of a fruit so that it does not bruise as easily during transport. Still others are being engineered for specific pharmaceutical or nutritional properties, such as "golden rice," which would provide supplemental levels of vitamin A.

Agricultural biotechnology has been rapidly adopted in the last decade, signaling industrial agriculture's continued growth. Since 1990, the USDA has approved approximately fifty genetically engineered foods or crops for commercial release, and nearly 55 percent of America's soybean crop, 35 percent of the corn crop, and almost half of the cotton crop are genetically engineered.[6] The total acreage planted in genetically engineered crops worldwide also reflects this growth. In 1996, the total global area planted with GMOs was 4 million acres; in 1999 it was 99 million acres.[7]

Supporters of GMOs claim that if "properly used, these new biotech crops have enormous potential to reduce pesticide use, improve human nutrition, and ease hunger in developing countries."[8] It is, however, hard to see the humanitarian potential knowing that the biotechnology industry is controlled by just a few corporations and that there is rapid consolidation of the genetic engineering, pesticide, and seed markets. Similar to the seeds of the Green Revolution, the new genetically engineered seeds, in order to reap the yield or production levels advertised, must be used in concert with pesticides, herbicides, and synthetic fertilizers. Thus, for the modern seed company to be competitive, they must own the company that develops the technology, produce the pesticide that works best with the new seed, be the distributor of the seeds, and so on.[9] This consolidation of the markets means that the potential of the technology is being controlled by corporate interests and profit incentive rather than governments or public research institutions. For example, Monsanto, Astra-Zeneca, DuPont, Novartis, and Aventis are the top five biotechnology corporations in the world. John Robbins, author of *The Food Revolution*, writes, "Together they account for nearly 100% of the market in genetically engineered seeds. They also account for 60% of the global pesticide market. And, thanks to a flurry of recent acquisitions, they now own 23% of the commercial seed market."[10] One of these companies, Monsanto, has created a line of crops that are

Fields That Dream

"genetically engineered to withstand repeated doses of Roundup, enabling farmers to spray their fields and kill weeds without killing the Roundup Ready crop."[11]

It is interesting to note that Dr. Kim Waddell, formerly a scientist with the National Academies of Science and study director for the Academies' report, *Environmental Effects of Transgenic Plants*, has stated that "Glyphosate/Round Up, as an herbicide, is one of the more benign chemical products available. Environmental impacts before and after Roundup Ready soybeans are difficult to compare, but I honestly think that genetically engineered crops have collectively reduced the overall use of pesticides/herbicides in agricultural production systems—which is ultimately good for the environment."[12]

But Waddell is quick to point out that, despite the overall reduction in herbicide use, the potential for the DNA of a genetically modified plant to contaminate the traditional seed supply is high. This "crop-to-crop gene flow" occurs in a variety of ways, including a "physical mixing of seeds or seed parts and pollen, which is carried by wind or insects to the female parts of plants and gives rise to new seeds."[13] In their study, *Gone to Seed: Transgenic Contaminants in the Traditional Seed Supply*, Margaret Mellon and Jane Rissler of the Union of Concerned Scientists reported that "seeds of traditional varieties of three major food crops: corn, soybeans, and canola ... are pervasively contaminated with low levels of DNA sequences originating in genetically engineered varieties of those crops."[14]

It is this danger of the genetic erosion of the world's seed supply, a rich history of seeds that have been developed over thousands of years, that evokes such pressing questions in the movement of genetic engineering. The potential fatal flaw of GMOs is that fewer and fewer companies are developing and maintaining seed varieties, thereby marketing a smaller cache of seeds to growing regions. These new "one-size-fits-all" seeds are less responsive to local growing conditions and preferences. The new method of developing and selling seed is in contrast to an era when a seed company only serviced a particular region, selling five to ten varieties of a crop based on specific soil types, water, and regional farming practices. The new biotechnology companies are now marketing five to ten varieties of crops for the entire country.

Since the turn of the century, the diversity of crops in America has dwindled dramatically. A study by the Rural Advancement Fund International (RAFI) found that by surveying nearly seventy-five types of vegetables, approximately 97 percent of the varieties given on a list made by the USDA in 1903 are now extinct. For example, in 1903, 307 varieties of sweet corn were grown in the United States. In the National Seed Storage Laboratory collection, there are only twelve different varieties, with a total loss of 96.1 percent.[15] With the addition of the Green Revolution, and now the biotechnology industry, the numbers are only falling.

Despite the innovations of the Green Revolution, people still went hungry because, even though the crops produced more food, consumers didn't have the money to buy them. And despite the worlds of opportunity in genetic engineering, there are frightening unknowns, uncertainties, and unresolved issues. Seed Savers Exchange, a nonprofit organization that saves and shares seeds writes:

> The vegetables and fruits currently being lost are the
> result of thousands of years of adaptation and selec-
> tion in diverse ecological niches around the world.

Fields That Dream

Each variety is genetically unique and has developed resistances to the diseases and pests with which it evolved. Plant breeders use the old varieties to breed resistance into modern crops that are constantly being attacked by rapidly evolving diseases and pests. Without these infusions of genetic diversity, food production is at risk from epidemics and infestations.[16]

This paradox is at the heart of the debate over GMOs. Monsanto says that GMO crops have the potential to eradicate vitamin deficiency and starvation in Third World countries. The company says small farmers worldwide will be able to produce a larger crop with fewer pesticides. And yet we also have evidence that GMO crops have the potential to bring on a catastrophic crop failure by causing dependence on a smaller gene pool of seeds. Small farmers are at risk of losing the integrity of their crops and, ultimately, they are at risk of losing their fundamental freedom to continue the legacy of a rich, safe, and diverse food supply.

▪ ▪ ▪ ▪

It is this freedom that Joanie and Michael cherish. I see them as a classic example of family farmers struggling against a threatening tide of biotechnology that is seeping into American agriculture. When Joanie said that all she wanted to do in high school was change the world, I wonder if she could have ever foreseen that changing the world would eventually mean trying to keep it the same. She and Michael want to continue the cycle of life that has endured for centuries; people making a living from the soil, saving seeds for generations, and feeding their communities.

Joanie and Michael take me on a quick tour of the farm before they head back to work, and they both emphasize how important this freedom is to them—that they have jobs where they are their own bosses and they are able to work outside.

Joanie smiles, "You have to feel pretty dang lucky when you get to be outside and watch eagles and trumpeter swans fly all over the place."

"We have snow geese too. They are pretty impressive." Michael scans the sky. "Hopefully they will come by before you leave. We have somewhere in the neighborhood of 50,000, and when they come over you can't even hear yourself talk, they are so thick."

The three of us turn around and head back out to my car where we find Rufus taking a snooze. Joanie hefts the dog aside so I can open the door, and we exchange our good-byes. She pulls two little zinnia starts out of her coat pocket and gives them to me. I am thrilled to have an addition to the other plant starts that are sitting on my windowsill waiting patiently for warmer weather.

I drive off the property searching the sky for any sign of even just one snow goose and think about how fitting it is that Rent's Due Ranch is on an island. No matter how inconspicuous the island is, Joanie and Michael are most definitely separated from the mainstream. They have told me that they are one of the only surviving farms in the area.

I cross the small bridge that takes me back onto the mainland and find myself wishing that this trickle of water separating farm from suburbia would suddenly go back to its natural state—that the technology of transforming island to solid ground, the technology of transforming seeds from fertile to sterile would reverse itself.

Fields That Dream

I close my eyes and imagine that 100 feet of water would suddenly come pouring forth from a wellspring and 50,000 snow geese would rise from the marshes with vast cries of delight.

Chapter Nine: Second Chances

It is estimated that there are nearly 300,000 homeless youth on the nation's streets every year, and Seattle's University District seems to have more than its fair share.[1] Many of the homeless teenagers hang out on corners and under business awnings, panhandling through curtains of matted hair and baggy black clothes. Locals, made uncomfortable by the sight of strung out fourteen-year-olds with red-rimmed eyes and mangy looking dogs, either ignore the kids or joke about how at the end of the day they count up the change they have collected and ride the bus home to the wealthy suburbs on the east side of Lake Washington.

I am also guilty of trying to cloak these young people in invisibility, imagining that they really are just cutting school and hanging out before going home to loving parents and a well-balanced meal. Dealing with the realities of what brought these youth to the streets—running away from physical or sexual abuse, financially unstable families, drug addiction, and neglect—seems too painful to recognize and impossible to mend with a couple of loose coins.

But Margaret Hauptman, founder and executive director of Seattle Youth Garden Works (SYGW), feels differently. In 1995, when her son turned sixteen, Margaret noticed a lack of alternative activities for teens in a city where gangs and homelessness were becoming a visible problem. She remembered her own experience of being in a garden and the healing she felt when working with the earth. "Gardens are beautiful, and they feed you. But for me, the garden was also a place where I learned a lot of lessons—lessons that were much easier to learn through nature, rather than looking at myself. I thought that I could do a lot of things at once here, and the wheels started turning: young people ... garden ... nature. And so, I just decided to do it."

Seattle Youth Garden Works is an employment training and education program that is structured around horticulture and business. It is open to all young people between the ages of fifteen and twenty-two who are homeless or at risk for drug abuse, academic failure, or any kind of criminal activity. Over the course of a year, fifty youth benefit from the program, with sessions holding up to ten participants at a time.

The youth are expected to treat the program as a job, and they are paid minimum wage. During the summer, SYGW has a stand at the University District Farmers' Market, and from early November through the holiday season, they make and sell wreaths. There is a short break, and then the program reconvenes for the Flower and Garden Show in February, where they have a display. After that, work in the greenhouse raising plant starts for wholesale to local nurseries begins.

■■■■

Margaret and I are talking in the kitchen down the hall from the large one-room SYGW office, which is perched on the top floor of a rickety old church directly across the street from the University District Farmers' Market site. It is toward the end of Margaret's workday, and her thick brown hair is slipping out of its bun, her green eyes heavy-lidded with the warm, late afternoon. She is explaining the unique nonlinear structure of SYGW that enables participants to jump into the program at any point. The six-month curriculum extensively covers horti-culture and other topics including environmental awareness, nutrition, life skills, and business.

She emphasizes, "The material is presented in a very immediate way so the participants can pick up and absorb it without a lot of academia or writing. Everything is very hands-on. For example, in our nutrition unit we have a nutritionist come in and teach a class that brings the young people some awareness about food, basic things such as reading labels and understanding what the fat and protein contents mean. But they are cooking while she is teaching, outlining different types of diets and explaining what it is that teenage bodies really need to grow healthy."

"And they get paid to go to class?" I ask.

"Yep," says Margaret happily, and then she elaborates. "We definitely emphasize the job-training aspects of the program and have realistic community standards as far as work and behavior on the job. But at the same time, they are getting an opportunity to learn about things that are important and why."

Margaret mentions that they have a strict policy on sub-stance abuse. "You can't be high at work, and if you have a problem where you are coming to work high—for example, the second time—then not only will you be excused for the day, you will have to speak with a drug and alcohol counselor before you are allowed to come back. We are not trying to tell them how to run their lives, but when it comes to work, you have to come up with a strategy."

I have often visited SYGW's stand at the farmers' market. They always have a nice variety of produce, and they also sell coffee, which, early on a Saturday morning, makes them very popular. Their displays are modest and artful, and everything is

Second Chances 125

sold with great enthusiasm. Customers can see the produce they will be buying as it grows in the SYGW plot at the University Heights Center community garden next to the market. They also have garden and greenhouse space at the Center for Urban Horticulture by the shores of Lake Washington, and Margaret is busy trying to secure land for another garden in south Seattle.

■ ■ ■ ■

Seattle Youth Garden Works is part of a larger movement that is working to bring gardening and sustainable agriculture into the heart of urban America. Thousands of school- and community-based garden projects are sprouting up across the country. Some were started by individual teachers and schools; other programs involve working with troubled teens or inmates at local prisons. Many are just collections of neighbors in highly urban areas carving gardens out of the cement.

With the help of organizations such as the Green Guerrillas, a group of volunteers that began turning vacant lots in New York City into gardens, there are now more than 1,200 community gardens in a city where more than 20,000 lots are still unused.[2] On the other coast, City Slicker Farm, a 4,000-square-foot farm in the middle of Oakland, California, teams up with the People's Grocery, a mobile natural food market that brings fresh produce and staple goods to one of the poorest urban neighborhoods in the country.

But pairing gardening with education is where communities are beginning to see real changes in their young people. Even state governments have gotten behind educational gardening initiatives. For example, to provide teachers with funding and garden-based curriculum that would be in accordance

Fields That Dream

with state educational standards, the California Department of Education launched the "Garden in Every School" program in 1995. The National Gardening Association estimates that it supports more than 10,000 educators nationwide in school and community programs that use gardens for educational purposes.[3]

Garden-based learning is not a new concept. Aarti Subramaniam, a research assistant at the 4-H Center for Youth Development at the University of California at Davis, notes that "by 1918 every state in America and every province in Canada had at least one school garden. In 1916 more than 1 million students contributed to the production of food during the war effort, following the proclamation of President Woodrow Wilson."[4] Interest in school gardens waned as teachers focused more on technology during the post–World War II era. There has only recently been a resurgence of school gardens, due to the birth of the environmental education movement of the 1980s and '90s.

Margaret, like most other urban and school garden coordinators, is adamant about teaching participants how to grow food organically. Though SYGW gardens are not certified, Margaret says, "It is crucial that one grows organically. It is more nutritious food, it creates a safer environment, and it is a sustainable practice. It is also a perfect analogy for what we are teaching about in the program: self-reliance and nature."

Then she laughs, "I don't know if you could be certified organic in an urban environment, what with all the buses and cars driving by! But it is easy for us to grow organically, certified or not. We have no slugs. We have almost no pests of any kind because we are surrounded by, well, asphalt!"

While educational gardens do focus largely on environmental awareness, they can provide excellent means for hands-on

Second Chances

training in math and science. A study conducted by twelve state education agencies found that students exposed to environmentally based educational programs performed better on standardized tests in reading, writing, math, social studies, and science. The study also found decreased discipline problems in classrooms, increased attention spans, and students who felt a greater pride and ownership of their accomplishments.[5]

Margaret acknowledges that a little bit of pride can go a long way in the building of self-esteem. "Virtually every week at the farmers' market we sell out of produce. The youth feel so much pride. It is a great feeling. They think, 'Wow, people want this stuff!'"

However, not everyone is eager when they start the program. Margaret laughs, "Some of them say, 'I don't want to get my hands dirty,' 'I don't want to bend over'! On the other hand, some are like, 'Do you have any ditches I can dig or weeds I can pull? I really like the physical work.' There are gardeners that love to tend the flowers, and some really love growing the vegetables. What I love to see, and it doesn't happen every day, is when they really get into it. They've gone from saying, 'I don't want to get my nails dirty' to knowing more about tomatoes than I do!"

Nicole Miller, a youth crewmember, already knew she liked gardening when she applied for a position with SYGW; she just didn't realize how much. She explains that while the structure of the job and needing to be somewhere and account to people has been helpful in providing her with a sense of stability and responsibility, it is the gardening that has become a true passion. "There is something really satisfying about planting something, watching it grow, and taking care of it. It sounds kind of strange, but gardening has gotten me to take better care of myself. I went from being homeless and eating anything I could get to being very environmentally aware and personally conscious of what I eat."

Nicole is a tall, strong young woman with bleached-blond hair and a burgeoning farmer's tan. She explains that her work with SYGW has solidified her goals for the future: she is going to go to a technical college and specialize in landscaping and garden design as soon as her SYGW position is completed. She speaks with incredulity about how much SYGW has done for her. "The staff at the program has been really good about not letting me stray on finishing my GED, and they are even helping me apply for scholarships and financial aid!"

She mentions that selling produce at the University District Farmers' Market is a way to bridge the gap between homeless youth and the community. "When people buy food from us at the market, I am sure they recognize the kids that are selling as the same ones that are standing out on the street. I think when people see us working it helps change the stereotypes about homeless youth. I mean, it is really rewarding to work for our money, to feel like we are not trying to leech off of people. We want to work. We are not just looking for handouts."

▪ ▪ ▪ ▪

Teenagers hear about Seattle Youth Garden Works in a variety of ways. Some find it by word of mouth, others through drop-in centers and transitional homes around the city. Half of the crewmembers are Caucasian, and the other half is made up of African Americans, Asian Americans, Latinos, and Native Americans.

Margaret runs the entire program with a staff of five, including herself. As she explains the configuration of her staff,

I notice how different she is from other nonprofit administrators I have met. She is younger than most in her position, in her mid-thirties. And when she tells me that she, too, was homeless as a teenager, I am surprised.

Margaret rests her elbow on the kitchen table and presses her chin into her hand. Matter-of-factly she says, "My family broke up when I was fifteen. I was left on my own to do things like finish high school. After leaving my home in California, I lived on a commune in Oregon for a year. Then I met some people who lived up in Seattle, and they invited me to move up here and go to an alternative high school, because at that point I couldn't have dealt with a traditional one. I ended up having a child really young. I know that my own background has a lot to do with me starting this organization."

She continues, "I remember feeling really stranded after high school, like 'What am I supposed to do now?' I didn't even know how to apply to college. I sent my SAT scores to Evergreen College in Olympia, but I didn't know how to go about getting financial aid or anything like that. There didn't seem to be anyone around to help me."

Margaret stops and looks at me. "On the one hand, I was very happy and grateful and satisfied at the freedom and experiences I had. On the other hand, more information is better and guidance is good, and I didn't have it! I think a lot of what my program is about is providing a place for these young and very often troubled people to be respected and given guidance. But not the heavy-handed, authoritative guidance they might be avoiding."

Margaret explains that much later she went on to community college and ended up with a graphic arts degree, working full time as a designer. She laughs, shaking her head slightly, "I

Fields That Dream

was doing a job that I thought I would really like, but then I didn't because I was sitting in front of a computer all day! I started thinking about what I really wanted to do. I knew that I loved gardening and I wanted to do some sort of social justice thing, so I decided to start Seattle Youth Garden Works."

I must look stunned at the thought of a single mother with no money starting a nonprofit organization, because Margaret laughs. "Yep, I did it! It gives me so much confidence ... I say, 'Wow, I did that!' I always like to share that story because I think, if I can do it, then you can do it."

Margaret's initial shyness has long since dissipated, and she talks openly about the roller-coaster ride of starting a non-profit. "It seems like everything is constantly changing. But it is always very challenging and interesting. I am a person who likes variety, and now I have got it! But it is a lot of work for not a lot of money."

She leans forward, raising her eyebrows, and whispers, "I have an entrepreneurial streak that has helped me with this program, and I am not sure where it will take me in the future!"

Work calls, so I thank Margaret for the time she has taken out of her busy schedule to see me. I head the opposite way down the hall and turn around in time to see her walking briskly back toward her office.

I am so impressed with Margaret and Seattle Youth Garden Works. Not only are they nurturing and saving the lives of the youth involved in the program, they are nurturing and saving the community at large. They are showing us that life and food and second chances can sprout through cracks in the sidewalk, and that compassion is not ignoring or denying a problem, it is doing whatever you can to help.

Second Chances

Chapter Ten: The Back Forty ...

Much to my surprise, Sunflower Farm is located in the Valley View Mobile Home Park in Bothell, Washington. Susan Wells told me this over the phone. When I couldn't quite figure how a successful sprout farm was based out of a mobile home, Susan laughed and said, "Well, bless your heart, you should come here and check it out."

So that is just what I am doing. It is February in Seattle, and it has been raining for days and days. I am driving north on Highway 405, which is on the east side of Lake Washington and the artery into the heart of computer land. I pass the cities of Redmond and Kirkland and start looking for Valley View Mobile Home Park.

The concept of the mobile home is as old as the first tribes of nomadic humans tens of thousands of years ago. From Gypsy caravans and American wagon trains to the rickety car trailers and smooth land yachts of the 1920s, mobile homes have been a fixture of human attention for centuries. It was during and after World War II, however, that the mobile home actually evolved into the idea of a long-term residence. According to Allan D. Wallis, a mobile home expert, there was a "sudden and tremendous demand for temporary portable housing created by World War II."[1]

During the war there were thousands of workers who needed housing, and after the war there was a desperate lack of housing for returning veterans, their new wives, and burgeoning families. Men building highways

133

and nuclear facilities after the war also needed inexpensive and easily constructed places to live. Mobile homes were the answer. Factories were already mobilized for mass production, and after the armistice, even war plane manufactures went into mobile home construction.

Efficiency was the name of the game. Mobile homes fit in perfectly with America's postwar mentality of a rapid and tidy expansion. Assembly-line homes could be conveniently transported on the new highways and deposited in the new suburbs. With war technology being applied to transportation, food, and housing, mobile home parks started appearing like rows of metal boxwood hedges.

Valley View Mobile Home Park is a slightly tattered reminder of these metal communities efficiently birthed in a flurry of necessity. The park, a conglomeration of modest mobile homes perched on a hill, feels refreshing compared to the sea of cookie-cutter luxury condos in the valley below. The namesake view is modest as well. It is a panorama of the freeway and its many veins, with a stretch of trees very far off in the distance.

Susan's mobile home fits in well with the others, except for the enormous pile of dirt that is sitting on a tarp in her front yard. I step around the pile, and Susan greets me at the front door with an extremely warm and toothy smile. Toby the dog, a cross between a border collie and something else, careens around Susan's body, barking frantically and nervously inspecting me.

Susan is tall with a light complexion and a rugged, kind face. Wisps of her light brown hair have escaped from under her polar fleece cap, which is pulled down to cover just the tops of her ears and skims the large frames of her glasses. She is bundled up against the February chill in warm layers. Susan keeps smiling as she leads me to the living room to drop off my stuff before the grand tour. The living room looks relatively normal, except for the fifty-pound bags of wheat seed leaning up against a large worktable.

"We have a pretty big operation here," Susan says proudly in her gravelly voice as she leads me out the back door. She laughs when my jaw drops in amazement. There is a huge greenhouse in what used to be the carport, and two other generously sized greenhouses have been placed strategically in the yard. Rows of white multileveled tables surrounding the greenhouses are stacked with flats and flats of seedling starts. It is quiet back here and oddly peaceful. I am instantly drawn to touch the plants, and I run my hand lightly over the surface of the broccoli sprouts. The young plants feel like a chattering creek sounds, gentle and light.

Susan smiles knowingly. "There is some kind of life energy in this backyard. Very few people walk by without running their hands over the wheatgrass. I think there is something about new life."

■ ■ ■ ■

A new life is what Susan had to create for herself to pursue farming. She begins to fill me in on the details as we walk back into the house and settle ourselves at the kitchen table.

Her voice is earnest and strong. "I worked for an insurance company in the claims department for about twenty-five years, and you just get to the point where you think, 'I don't want to do it anymore,' and you just jump off that bridge, you know? It was scary. But my children were grown. I had to decide [that] if the farm didn't work and I had to live on the street, so be it. That is how desperately I wanted out of the insurance industry. And so, I made that decision to take a chance. If it didn't work,

The Back Forty ... Inches

135

well, you don't want to die thinking, 'I should have done this or that.'"

Susan and her business partner, Eric Alexander, started the farm in 1991. "Eric was putzing around with his little sunflower sprouts and growing them for his own use, and I had experimented with growing wheatgrass for juice. We had a little money saved, and I knew we could get by for a year and build that greenhouse. And that was the beginning of it. We planted our sunflowers, and I called the local grocery stores and told them the idea that I had, which was to sell sprouts that were alive. You see, all of our sprouts are delivered growing in little containers instead of snipped, so they are very, very fresh. Local natural food stores were very encouraging, and we would go around to different establishments and restaurants carrying our little sprouts and asking people to taste them."

Susan sits back and shrugs. "Eventually the business just grew and grew, and it seemed to be synergistic. I suppose fear is a great motivator. It had me up until two or three o'clock in the morning picking the hulls off sunflower sprouts, cleaning, drying, sorting, and arranging edible flowers in the containers just so."

The business is indeed doing well. They sell to a number of natural food stores, juice shops, and some of Seattle's finest restaurants. Sunflower Farm plows through 300 pounds of wheat seed a week, not counting the seed for all of the other sprouts they grow, which include sunflower, broccoli, arugula, cilantro, and pea. Another sign that the business has grown is that Susan and Eric now have three part-time employees.

Susan explains to me in a serious tone, "When an article about broccoli sprouts preventing cancer came out, we couldn't handle all the business by ourselves anymore. Eric and I just called the high school and asked if anyone who was interested in horticulture wanted to come out. They sent out a fellow, and he has since graduated and gone on. But before he left we asked him if there would be anyone else that wanted to come, and that's exactly how we got Nick and Adam."

Susan pauses for breath and then smiles expansively. "I hope you get to meet them. I love them so much I would adopt them! If I had a big house they'd all be living with me if they wanted to!"

Susan and Eric give the boys a lot of work and pay them well for it, a very respectable nine dollars an hour. "This year I am letting them handle the Pike Place Market. I feel real comfortable letting them do it. It gives them a real sense of responsibility. I feel good about what we pay them. And that is what this is all about, feeling good. We know that we will never be rich, but I'm content to be here, doing what I love."

Susan lets out a hearty, raspy laugh, points out the window, and says, "I'm happy here. I mean, look at the view!"

Toby barks at Susan's sudden burst of laughter, and it takes us a few minutes to calm him down. She goes back to talking about the boys. "You know, Nick was working at McDonald's before he started here. I guess he just took off from McDonald's one day and showed up here with that first guy that we hired. That's what they do, they show up. We asked Nick if he wanted to come back the next day, and I guess he never did go back to McDonald's."

Susan remembers that at one point they hit some financial problems and couldn't give Nick the raise they wanted to. When they told him the bad news, apparently all Nick said was, "Just as long as I can come back. Just as long as I can come back!"

"It melts your heart," Susan sighs.

■■■■

Obviously it is unusual for two suburban boys to have the opportunity to work on an organic farm, albeit an unconventional one. But in a country where food companies spend $12.7 billion a year on marketing to children,[2] of whom 15 percent between the ages of six and nineteen are overweight,[3] any chance to get kids outside and active is a good one.

Fast food is one of the largest culprits of childhood obesity. With the help of marketing agencies specifically geared toward children, fast-food companies have targeted young people through "playlands," Internet clubs, and partnerships with film companies that promote their movies through toy giveaways.[4] In part, because of such marketing strategies, childhood obesity has reached epic proportions. Marion Nestle, a nutritionist, reports, "Rates of obesity are now so high among American children that many exhibit metabolic abnormalities formerly seen only in adults," including problems such as adult-onset diabetes, high blood pressure, and high cholesterol.[5]

Concerned citizens are working to sue fast-food corporations through class-action lawsuits, introduce legislation that targets obesity, and remove junk-food vending machines from public schools. But others are focusing directly on educating children about nutrition, the benefits of physical activity, and where their food is coming from.

For example, in Goleta, California, completely surrounded by suburbs, there is a twelve-acre farm run by Michael Ableman, a farmer, writer, and photographer. Ableman started an environmental education program on the farm, and he once had twenty-five kids stay with him for five days.

The first night there were a number of cries of "I wanna go

home." By the end of the second day, home was a distant memory replaced by thoughts of strawberries, goats, and the wonders of earthworms and living soil. Parents were surprised to find that their kids were too occupied to acknowledge them. They were even more surprised to see their sons or daughters scarfing down fresh vegetables at mealtime.[6]

These are city kids—flabbergasted to see milk actually come out of a cow, not a carton, and bowled over to witness a carrot being pulled from the ground. Like the kids on Ableman's farm, Susan's teenage employees love the wonderment and physicality of being outside.

She smiles, "They really like just being kids. You know, being in the dirt. In the wintertime it is challenging, but in the summer they are outside all the time."

Susan laughs as she thinks of another perk of the job. "Oh, and I feed them! A lot of times I'll feed them before they go home."

There is a sudden rustling outside, and Toby starts barking and wagging his tail so hard that he looks as flexible as a slinky. "Oh, the boys must be here," says Susan, getting up from the table to let them in. She opens the door, and two teenage boys amble in, petting Toby who is now frantic with excitement.

"Hey Sue. Down Toby," says one of the young men.

Susan introduces me to Nick and Adam, who, for the life of me, I can't tell apart. They are almost identical in their designer sweatshirts, big, baggy jeans, and baseball caps with the brims expertly squinched down to form tight, round curves.

"Hey. Nice to meet you," says one, slightly sheepish but genuine.

"I am going to make some coffee," says Susan. "Why don't you boys tell Jenny a bit about what you do around here."

They seem interested in talking with me, though a bit shy. I find out that they are both high school seniors and they have worked at the farm for a year, about eleven to fourteen hours a week. After graduation one plans to apprentice for an electrician, the other will go to community college and plans to continue on at Sunflower Farm. Their sweetness and earnestness comes easily to the surface when they speak.

Nick/Adam says, "We got lucky to be here. It's different. We are actually doing something that helps people out. Growing stuff, it's cool."

The other one interjects, "And we have good bosses. Both of them are really nice. I don't know any other bosses that would make us dinner!"

I ask if either of them thinks about going into farming. One answers, "I don't know what I want to go into. Maybe business. It's cool being part of this small business and watching it grow."

They talk willingly about their favorite parts of the job and what sprouts they like to eat the best. The boys actually drink wheatgrass juice and make smoothies out of the sprouts.

Nick/Adam says, "I think that's one of the reasons why this business goes, because they don't care about just making all the money. They want to be sure the product is what people want. They make sure their product is the best that they can put out."

Susan comes in with a steaming cup of coffee and offers some all around. She smiles proudly at the boys. I tell them all how much I admire what the farm does and how happy they all seem. Susan settles herself at the table and then says, "I think, Jenny, I think if you do what your heart tells you to do, as far as making a living, I think it will work. I really do. If you are tenacious enough to step down a little as far as finances are concerned, I think eventually you can do what you want to do."

Fields That Dream

We sit around the table for a little while longer, and the boys check in with Susan about what they will be doing for the day. Toby snoozes on the floor by Susan's feet, and the rain makes a melodic pinging noise as it hits the metal roof of the mobile home.

I find that there is a subtle revolution happening at the Valley View Mobile Home Park. No one would know that in this inconspicuous mobile home Susan Wells is flying in the face of American efficiency, shepherding young souls from the dangers of fast food, and leading them to the freedom of playing in the dirt, growing pure, healthy food, and running the palms of their hands over soft, new life.

The Back Forty ... Inches

Chapter Eleven: Sky's the Limit

It is the very beginning of April, and there is still snow on the sides of the road that leads to Full Circle Farm. The farm is in North Bend, the last major town before Snoqualmie Pass, which is the gateway to eastern Washington. Surrounded on all sides by forest and half an hour from Seattle, it feels as if North Bend is in a gawky adolescence, no longer a small logging town and not quite a full-fledged suburb.

Full Circle Farm is located a few minutes from the downtown area and fits into the neighborhood well. There are maybe ten houses on the whole street. Toward the end of the block, the houses are only on one side, making room for the base of a large slope that seems to be the little cousin of the mountain behind it. Each house is surrounded by fields where horses wearing thick wool blankets graze and plastic swing sets and SUVs rest in the driveways, covered by a thick layer of frost.

On the property there is a tumbledown ranch house that operates as Full Circle's office. The fields, which begin as soon as the house ends, are exploding with bright chartreuse growth: the lettuce seedlings and tiny vegetable starts are almost florescent in their newness. A row of sheds on one side of the fields houses all of the equipment, and greenhouses on the other stand erect and ghostlike in the shifting morning light.

As I pull into the driveway, a number of guys are milling around a large trailer that looks as though it has just been delivered. When I get out of the car, I hear a heated discussion in Spanish about how to set the trailer on the new foundation. In my halting Spanish I say, "Hi, I am here to see Andrew Stout. Do you know where he is?"

The men break into wide grins at my feeble attempt at communication. One of them steps forward and says in perfect English, "Hi, I'm Erick. I think he is around back, let me go and get him."

After a few minutes, Erick returns with a tall, wiry young man whose fair skin is shockingly bright compared to the brown skin of his crew. Andrew and I greet each other, and he fills me in on the history of his farm.

"My partner, Wendy, and I moved out here from the Midwest in the fall of 1995, and I planted my first crop that following spring. Full Circle is about fifteen acres, and we cultivate nearly every inch of it. Here on my farm I don't grow anything but greens. I got rid of growing tomatoes, potatoes, and other things that weren't making any money."

Andrew shakes his head and seems to chuckle at his own ignorance. "Three years ago, you name it, I was growing it. I started growing it all, and then every year I get rid of the stuff that doesn't make the money. I still grow beets, carrots, and onions, but it is mostly greens: salad mixes, baby greens, collards, lettuces, chards, mustards, mizuna, arugula, baby bok choy, all different kinds of cabbages. I really try and focus on specialty items. I mean, there are already so many tomato and potato farmers around, and for a small farmer to make it these days, it is most lucrative to find a niche for yourself."

Full Circle Farm started out with a stand at the U-District Farmers' Market and a small community supported agriculture program (CSA), which Andrew has been trying to double every year. This year he hopes to have 150 members. The CSA program invites people to pay a seasonal subscription fee to receive a large basket of produce every week. Wendy runs the pickup sites and customer service end of the CSA program and also plans and implements the flower segment of the farm. Her main focus, however, is acupuncture school.

Andrew laughs and says almost gleefully, "So, I get to do all of the dirty work!"

It is obvious that Andrew not only enjoys his job, he is passionate about it. It is nice to see his enthusiasm. Since Andrew is only thirty, he has a long row to hoe. But he definitely has a plan.

He tells me fervently that he is working on building the business. In addition to the market and CSA, Andrew sells his greens to a number of the premier restaurants in Seattle, many of which pay tribute to the farm on their menus. But his real pride and joy is the distribution company he has started.

"You see, I don't just sell my greens to the restaurants. I market everything. I have an extensive price list: heirloom tomatoes, beans, corn, melon, squash, beets, you name it. If it is organic and grown in Washington, I can get my hands on it!"

Andrew explains that Full Circle Distribution is the beginnings of a cooperative, and he has already hired a buyer and drivers. He got a grant from the King County Agricultural Commission to help start it, and now he picks up food from about twenty farmers from around the state.

Andrew tells me his motivation for starting the business: "You are never going to get rich dirt farming, you have to supplement. I started the farm as a base that runs itself and does reasonably well. Moreover, it has good philosophical, social, and environmental roots. Now I am trying to find a business that makes some money."

Sky's the Limit 145

"There are a lot of benefits to a statewide distribution company," he continues. "You can put a name to the food, which is really what we are about. I send an interviewer and photographer out to all of the farms that I work with so I can get biographies together and send them out to the stores, putting a face to the carrots. This distribution company strives to connect farms together, to pass information along. I want this distribution company to challenge the local competition. I want to see myself with thirty trucks on the road supporting small organic farmers in every little nook and cranny of Washington. My learning curve is steep, but I am feeling hopeful about it."

Andrew leans forward and says earnestly, "I am a young farmer. I started this at twenty-six. Other young farmers like myself have to have other jobs to be able to make a living. But for so many other farmers, this is their second career. They already own their houses, they have already put their kids through school, and they are already pulling a pension."

"But," he continues emphatically, "this is what I do!"

Andrew chuckles and shakes his head in faux disbelief. "I just went down to a fifth-grade class for their 'Career Awareness Day.' I said, 'Yes, you too could become a dirt farmer!' But seriously, I told them that there is so much that an organic farmer can accomplish, like supporting the landscape and providing a healthy and nutritious food source."

Andrew doesn't see an alternative to organic farming. "I guess I was just raised with it. My grandfather was a master gardener, my grandmother was a naturalist, and my parents gardened too. It was the kind of deal where we would go up smelt fishing and then actually plant your beans and put a fish in the hole with them, just like the Indians did."

Andrew didn't exactly know that he wanted to farm from the start. After college, he and Wendy moved out to Bend, Oregon, where he says they enjoyed "poverty with a view." While Wendy was out raising whooping cranes for release into the wild, Andrew was working as a manager for a local landscaping company.

Andrew takes off his baseball cap to run a hand through his reddish brown hair. He recalls without much enthusiasm, "I would drive around town, mow people's lawns, put in plants, and all sorts of other super easy stuff. It wasn't organic because it wasn't my company, and the boss would tell us to go spray. After a while I started thinking, 'What the heck am I going to do with my life?' Neither Wendy or I were attached to Bend, so we moved back home, and I thought, 'Farming! That's noble!'"

I am curious why Andrew and Wendy chose to relocate to Seattle. Andrew replies without hesitation, "We wanted to have the farm within sixty miles of an urban center, and Seattle is fantastic! The timing worked out really well. When we got here there were so many good restaurants on the scene, and the local farm base for who was supplying the restaurants had never been thoroughly developed. It is not like San Francisco where there are 700 farms vying for all of the business in the city. It is a lot easier up here because there is so much room for anybody that wants to get into the market. There is a great political climate here as well. People want to see the farms succeed, they want them to be part of the community."

I certainly agree with everything that Andrew is saying, and I know to a certain extent that he is right about people wanting the farms to be successful. But there is a missing link in the chain. Andrew tells me that without a doubt he is going to have to start looking for land in the near future because he leases the property that Full Circle is on, and he is certain that the acreage will be developed. This echoes the stories of urban encroachment that so many other farmers have told me.

What I don't understand is this: I know there is an affordable-housing crisis in the greater Seattle area. I know that people need convenient places to shop and open spaces for their children to play once they live in these new suburbs. But where do they expect these farms to go? Soccer fields and housing developments are coming at an increasingly dangerous cost.

Andrew speculates optimistically, "I am shopping for land all the time, and hoping to buy in Duvall or Carnation. I need to spread my wings a little more. We grow a boatload of food out here, and if I had a little more acreage, it would make it a lot easier to do that."

■ ■ ▨ ▧

Andrew and I decide to get up and stretch our legs, and he takes me on a tour of the farm. As we round the corner of the house, I ask if this is where he and Wendy live.

"Oh, no!" He seems amused by the idea. "We don't live in North Bend, we live on Capitol Hill."

Capitol Hill is one of the hippest, most urban areas of Seattle, and Andrew laughs at my look of surprise. "I love the city. Living there and working here provides a good balance for me. I don't burn out in the city and I don't redneck out here."

He explains that the house is used as living quarters for some of the seasonal staff. During the height of the season he employs up to twenty people. Erick is the foreman of the crew, and Eviva is his right-hand woman. He also has five drivers and at least ten field hands. Andrew tells me that his crew is generally half Mexican and half Caucasian, and he likes the interracial balance. What he finds particularly interesting is the gender balance.

Fields That Dream

"Last year," he tells me, "it was mostly women. This year the crew will be mostly men. You never know what the next bit of crew will be like."

Andrew explains that while Full Circle is one of the larger organic farms in the Seattle area, it is miniscule compared to some of the organic "giants" in California and Mexico.

He says that the size of organic farms that have emerged over the last fifteen years is drastically varied—from the very small family-run businesses to the "industrial organic" farms primarily owned by large corporations. His farm falls somewhere in the middle.

The rise of corporate organic farming began when sales in organic food started to grow exponentially in the early '90s. The United States Department of Agriculture (USDA) responded with the call to create legislation to regulate organic production. Mainstream food giants such as Heinz, Dole, ConAgra, and Archer Daniels Midland (ADM) began to hustle to get into the organics market.[1] The introduction of these transnational corporations into organics has significantly transformed the world of sustainable agriculture. No longer a network of small and midsized farms selling in a local market, organic farms have been turned into serious business, which depends just as much on shipping, storage, and processing as conventional food manufacturers do. Writer Michael Pollan notes that "the biggest organic operations in the state [California] today are owned and operated by conventional mega-farms. The same farmer who is applying toxic fumigants to sterilize the soil in one field is in the next field applying compost to nurture the soil's natural fertility."[2]

While it is a triumph to the sustainable agriculture movement to disprove the old myth that large-scale organic production is not possible, for many it is disconcerting to discover that the organic foods available in the supermarket are produced by giants such as General Mills. Furthermore, there are no visible signs that these corporations own the organic companies. For example, on a box of Cascadian Farms "Purely O's," an organic version of Cheerios, there is no mention of General Mills, only of Small Planet Foods. But General Mills bought out Small Planet Foods in 1999.

This invisibility of corporate ownership of smaller companies is very common, and it provides the market with the illusion of diversity. However, diversity is truly an illusion. "In the United States alone, just three companies—Philip Morris (Kraft Foods, Miller Brewing), ConAgra, and RJR-Nabisco—accounted for nearly 20% of all food expenditures in 1997."[3]

Not surprisingly, these corporation-backed industrial organic farms are now pushing small farms out of business. With mass production comes the ability to lower prices, and while this may make organic produce available to a wider population, it is destroying the niche market that has enabled small organic farmers to survive. Thus, Pollan laments, "many of the small farmers present at the creation of organic agriculture today find themselves struggling to compete against the larger players, as the familiar, dismal history of American agriculture begins to repeat itself in the organic sector."[4] This consolidation of agriculture continues to encourage the development of farmland into suburbs, for who needs small farms dotting the country when they can all be lumped together for the sake of efficiency?

Unmentioned in the discussion of industrial versus small-scale organic is the question of fair labor practices. Although not necessarily inherent in the world of small-scale organic agriculture, I like to believe that workers are afforded better

Fields That Dream

treatment on family farms, and I express my concern to Andrew that industrial organic farms may be treating their workers just as poorly as conventional ones—minus the pesticides.

Andrew relates his opinion as we walk out toward the fields and eventually come to rest by leaning against one of the tractors. "The food system in this country is not set up to pay agricultural workers a fair wage. I read an article that said this country pays only 14 percent of its income, if that, on food. We are one of the few countries in the world that doesn't pay at least half. Most Americans don't care about food, they don't see beyond the cash register. Until they are willing to pay a little bit more for their food, working conditions on large farms are not going to improve. But there is a big difference between corporate- and family-owned farms. This spring Erick, Eviva, and I went down to California to tour organic farms. A lot of those Latino guys on those farms have been working on the same farm for seven to ten years. They make fine money, have total responsibility, and essentially run the farms. But when it leaves the family and goes into the corporation, that is where the dividing line is."

Andrew feels certain that he is running his place like a family farm. "I want to pay my crew well. If they are happy and work-ing well, then everything that comes off of the farm looks good. It translates right down the line. I want to see that the farm runs efficiently. That is what our whole focus is. We get the necessary pieces of equipment, we get the necessary farming styles and practices, and then we hope that this place runs flawlessly."

Andrew believes that while small, self-sustainable farms have their place, they also have a downside. "On those small farms," he maintains, "the farmer wears forty hats. He or she is the marketer, the mechanic, the accountant, and, not to mention,

151

Sky's the Limit

the grower. My belief is different from a lot of farmers. Ironically, I don't think that is 100 percent sustainable. On those farms you are an island unto yourself, and you are only as good as yourself. What happens to your farm if you break an arm?"

Andrew is ready to get back to work, and so I suggest that we walk back toward my car. He stops to pull out a few weeds on our way. I ask Andrew if he feels different from peers he meets in the city.

He answers immediately, "Oh yeah, I love it! I am doing exactly what I want to be doing. I am in control of my destiny. I mean, if somebody asks me what I do for a living and I say that I am an organic farmer, it is almost incomprehensible to them. I ask people, 'Don't you do what you want to do?' A lot of people I meet don't."

Andrew pauses, as though mourning their loss, and then says, "I feel good. Every day I wake up and, even if they are not great days, I have an underlying sense of accomplishment. Whether the farm and distribution company make it or doesn't make it, going there was well worth it. It has been fun."

As I drive back to Seattle, I think about what an interesting combination of ideals Andrew holds. He does not let his business savvy or his ambition sway his commitment to organics or small-scale agriculture. In fact, it looks as though he is combining the two in a promising partnership. I wonder if Andrew and his Full Circle Distribution Company are a bridge to the future, a compromise between big business and small farming.

I know that Andrew Stout won't ever be able to give General Mills or Philip Morris a run for their money. But I do wonder if he will provide new meaning to capitalism in the new millennium, where the backs of others are not broken when climbing to the top. He does seem to be off to a good

Fields That Dream

start. Perhaps someday he will have a franchise in every state, distributing local organic produce to local people.

Coming from the east, the view of Seattle is breathtaking. The initially icy, gray morning has turned into a gorgeous, blue day, and the sun is glinting off the skyline as I cross the I-90 bridge that leads straight into downtown. Andrew has definitely landed in a city with American Dreams ripe for the picking. I remember him saying, "Who knows where the future will send Full Circle Farm? Value-added products, franchise distribution ... you name it. If I maintain the farm well, then anything that flows from it will have a firm basis to build on. I really think the sky is the limit."

From my vantage point looking over downtown Seattle, the frosty blue waters of the Puget Sound below and the Olympic Mountains towering above, the sky *is* limitless.

Chapter Twelve: Grafting

I have just driven past a sign that says, "Welcome to Yakima, the Palm Springs of Washington." I don't think this is a joke. I look around, but from the highway all I can see are rows of boxy chain stores, truck yards where various limbs of semis lay stranded in disrepair, mobile home parks, and homespun signs for local fruit stands. It is a blearily cold day in the middle of March, and the low-lying hills at the lip of the wide valley are still covered in the shocking gold of dead, dry grass. Even in a couple of months, when the grass is green and the orchards are blanketed in blossoms, I can't imagine this resembling anything close to Palm Springs, and for that I am thankful.

I am headed to Mair Farm-Taki, which is located outside of the small town of Wapato. Wapato, on the eastern edge of the Yakama Indian Reservation, is a

stone's throw from the midsized city and agricultural hub of Yakima. The town is worn around the edges but picturesque, with old clapboard buildings and tree-lined streets. There is a Mexican grocery or bakery on nearly every block, an indication of the rapidly expanding population that has settled in the area to work in agriculture.

I drive down a bumpy dirt road that turns into the Takis' driveway. The land is quiet and the air still when I stop the car, and I am immediately struck by the profound peace, almost gentleness, of the farm. At the center is a pond, bordered on one side by a bank of trees that lead off into wilderness and on the other by fields that roll into orchards. A log bridge leads to a small island on the pond, where a lone pine stands like a sentinel. The house, barn, and outbuildings are nestled in a corner near the driveway.

Katsumi has come out to greet me, and his brown, handsome face is covered with a smile so large that it crinkles lines around his almond-shaped eyes. He ushers me out of the cold and into the 1950s-style ranch house, where I slip out of my shoes and don a pair of black slippers. Ryoko emerges from the kitchen, wiping her hands on her slacks. She has shoulder-length salt-and-pepper hair, large brown eyes, rich wheat-colored skin, and a delicate smile. Together she and Katsumi make a stunning couple.

We retire to the living room, a wacky but elegant balance between vintage 1950s furniture and Japanese art and pottery. The seating faces a picture window that looks out on a stone garden surrounded by rosebushes. Ryoko brings in a steaming pot of twig tea, and we all sit down to talk. Both of them speak with thick Japanese accents, as though their mouths are still unaccustomed to the harsh consonants of English.

Fields That Dream

Ryoko explains that the place has undergone quite a trans-
formation in the five years that they have lived on the property.
"We first came here to help Rose Mair, the elderly woman that
owned the place. The farm was like an overgrown jungle, and
the house was dark. The gate was always closed. But I loved this
place from the beginning."

After Rose passed away, the acreage was going to be claimed
by the State Fish and Wildlife Department. Interested in saving
the land as a refuge, the Takis were also committed to upholding the property's farming legacy, so they decided to leave half
of the land to natural habitat and farm the remaining acreage.

Katsumi shakes his head. "The place was so out of care. It
had just been left. We had to do so much work to get the farm
going again."

Ryoko agrees, "There is still so much work to do. However,
it wasn't so hard to transition to organic, because the place had
been untended for so long."

I ask about the name, Mair Farm-Taki. Ryoko says quietly,
"Since Rose and her husband had no children, their name would
have been lost. This is the way that we can respect them."

Katsumi points out the window. "The stone garden and
rosebushes are in her memory. The Mairs planted the roses,
and since she loved rocks so much, we collected them in one
place to form a rock garden."

Ryoko reaches over to a couple of framed pictures on the
end table. The first one was taken under the maple tree out
front. It is of a Caucasian couple dressed in plaid and polyester,
in a grainy color finish typical of the 1970s. They are holding
hands and smiling. The next photo is of Ryoko and Katsumi,
standing under the same tree. The color is crisp and clear, and
auburn sunlight filters through the trees. Their arms are

draped around each other, and they are beaming.

"You see," says Ryoko, "This is the Mairs and now this is us. We will carry on their name and their memory. We are so thankful to them, and we are so happy here."

The Takis have restored the fifteen acres of orchards, and the fields are now thriving with specialty vegetables, some of them Japanese varieties of radishes, tomatoes, and greens. Fruit is their specialty, however, and they grow many varieties of apples and stone fruit. During the summer, theirs is the first stand I head to at the market. I want to make sure I get the cherries, small Japanese plums, and peaches before they sell out.

▰ ▰ ▰ ▰

Turning this dormant land into a vibrant and flourishing farm fits right into the history of farming in eastern Washington. The arid eastern side of the Cascade Mountains wasn't always farming country. Officially, the region is still considered high desert, but it has been richly irrigated by the rerouting of the Columbia River.

In the 1940s the Roosevelt and Truman administrations financed the building of the Grand Coulee Dam, then the largest man-made structure in history.[1] Harnessing the power of the Columbia River produced the cheapest electricity in the world and was instrumental in America's war effort. The energy was wired to Boeing, which manufactured war planes, and to the Department of Energy's Hanford Plant, which made plutonium for the atomic bomb dropped on Nagasaki in 1945.[2]

After the end of World War II, when the need for such massive amounts of energy for industry receded, the water was pumped into a valley, subsequently renamed Banks Lake. The lake has a network of canals flowing out of it, providing the

Fields That Dream

cheapest irrigation water in the nation.[3] As a result, agriculture is now the economic backbone of eastern Washington; wheat, corn, potatoes, peas, and apples, and an extensive network of farmers, agricultural scientists, food processors, and fertilizer companies are all found in the area surrounding Yakima.

But transforming desert to farmland means having to literally make dirt. Although the desert produces natural topsoil, it is sparse and silty. To create topsoil well suited for agriculture, one needs to add hefty amounts of water and organic matter to the ground, which is what the farmers of eastern Washington did once the floodgates of the Columbia were opened. In the scheme of things, the topsoil of eastern Washington is in its infancy, since it generally takes between 200 and a 1,000 years to create an inch of topsoil.[4]

Topsoil is an absolute necessity for agriculture—indeed, for human life on earth. But due to wind, human-caused water erosion, and pollution, topsoil is disappearing at an alarming rate. Since the 1950s, the world has lost nearly one-fifth of its topsoil, and the soil "lost in the United States alone, if loaded into freight cars, would make a train that would encircle the planet twenty-four times."[5]

It is hard to say exactly what the level of pollution is in the topsoil of eastern Washington. It is a known fact that there are lead-arsenic and DDT residues in the soil of old orchards. To control coddling moths in the years before World War II, apple farmers sprayed lead-arsenic onto their trees. After the war, it was DDT. Nowadays, testing for both lead-arsenic and chlorinated hydrocarbons (DDT) is routine in eastern Washington orchards that have been around for more than fifty years. Often there are low levels of pesticide residues in the soil, but nothing that violates national organic or conventional growing standards.

Grafting

If residue is found, the type of crop grown may be regulated, since root vegetables such as potatoes and carrots are more susceptible to contamination than tree crops because of their proximity to the soil.

But the effects of Hanford on the soil are yet to be studied. In fact, it seems as though there is a big gaping hole when trying to connect the history of Hanford to the current state of the soil in eastern Washington. The plutonium that Hanford manufactured during World War II produced a number of byproducts, including iodine-131 and plutonium-239. Both were intentionally released during routine operations throughout the 1940s and '50s in quantities 500 times greater than in 1979's nuclear disaster at Three Mile Island.[6] The iodine-131 and plutonium-239 were carried for hundreds of miles on the wind and settled everywhere, including the wheat fields all over eastern Washington. They were then consumed by cattle, thereby polluting the topsoil, the milk, and the beef.

However, pollution from Hanford is not a thing of the past. While iodine-131 stabilizes into nonradioactive xenon-131 relatively quickly, it takes plutonium-239 24,000 years to stabilize.[7] Moreover, there are documented spills from the 1950s and outdated radioactive waste tanks that are known to currently be leaking tons of uranium into the soil around Hanford. That waste is slowly seeping into the groundwater and moving toward the Columbia River.[8]

Although documentation of these intentional releases and spills of radioactive waste exists, there is little if any scientific data tracking their effects on the people, land, or agriculture of the region. Large debates in the scientific community on the leaks' relevance to agricultural production and public health in general continue to take place. There are no routine soil tests

Fields That Dream

for plutonium-239, and there is little to no discussion of how a polluted Columbia River will continue to irrigate crops in eastern Washington and Oregon.

The environmental unknowns and subsequent lack of scientific data on Hanford and the pollution of the Columbia River remind me of the aboveground atomic testing in the Nevada desert during the 1950s. Terry Tempest Williams, author of the memoir *Refuge*, writes about the effects of nuclear fallout on her family and the wildlife of Utah. "Again and again, the American public was told by its government, in spite of burns, blisters, and nausea, 'It has been found that the tests may be conducted with adequate assurance of safety under conditions prevailing at the bombing reservations.'"[9]

Although the evidence of soil contamination and radioactive pollution in eastern Washington is not as well known as it is in Nevada and Utah, it is equally insidious, and the silence surrounding it is just as dangerous. Eastern Washington soil grows apples and wheat that feed the world, thereby making this newly productive desert soil of global concern.

And still there is silence. During their organic certification, Ryoko and Katsumi weren't required to do any soil testing. Neither were any of the other organic farmers they know. Perhaps there is no need for worry, or perhaps looking for pollution from Hanford in the soil is like searching for a needle in a haystack. Dr. Bob Stevens, an agricultural scientist at Washington State University's Prosser Irrigated Agricultural Research and Extension Center, says, "Heavy metals are extremely hard to test for in soil. Everyone has his or her own perceived risk level. For me, on a scale of 1 to 100, I would say that plutonium-239 is around a 2 or a 3. I am much more concerned with excess nitrogen and phosphorus in the groundwater." He continues, "Soil does have a legacy. We often expect soil to heal itself, but it doesn't. However, our technology has advanced so far that we can now take better care of it."[10]

Farmers like Ryoko and Katsumi are working to create a new legacy for the soil, but often they feel isolated in their efforts from the rest of the farming community in the Yakima area, which is slow to change.

Ryoko observes that, although there is a lot of competition at the Yakima Market, they are beginning to win over some of the locals. "In the beginning we didn't want to go there, especially as an organic farmer."

Katsumi adds, "In Yakima they use a lot of pesticides."

Ryoko concurs, "This area isn't really ready for change. Many people just want big quantity for little money, and we charge a little more than other farmers. But the market manager was so enthusiastic. I don't know how many times he called here! 'Just try it,' he pleaded. We couldn't resist his efforts, so we gave it a try. We sold a lot of apricots in Yakima, and that was a surprise. Most people didn't care if it was organic or not organic, but we gave away so many samples and they liked the taste of them."

But their reception at Seattle markets is completely different—people have come to expect organic fruit. However, Ryoko explains that the University District and Pike Place Markets are the only Seattle markets Katsumi goes to. They are trying to transition to selling only at the University District and Yakima Markets because selling in Seattle is an exhausting three-day project of harvesting, selling, and then recuperating.

In addition to the Seattle and Yakima markets, the Takis have a small community supported agriculture program of

Fields That Dream

eighteen members. Ryoko is also interested in starting some side businesses. She wants to open a roadside fruit stand where they sell the farm's organic produce, organic grains, and dry goods. They are also looking into renting the property for outdoor weddings and events.

It is imperative that the Takis make enough money, as neither Katsumi nor Ryoko have been granted American citizenship, and if they don't make enough income, their residency may be jeopardized.

Katsumi and Ryoko have been experiencing hardship regarding their status in the United States. However, they tell me that the community has been extremely supportive. Their church started a letter-writing campaign to the senator and local congressmen, and the town newspaper had a front-page article documenting their fight for citizenship. Ryoko tells me she was overwhelmed by the efforts of the town and expresses her gratitude by being as involved in the community as possible.

When Ryoko heads to the kitchen to make more tea, Katsumi begins to tell me more about his background. Raised in the outskirts of Kyoto, he was the one Christian in a family of Buddhists. He left home to study agriculture and fisheries science in college. After graduation, he worked in a government job for two years and then moved to Okinawa, where he met Ryoko.

Katsumi says, "I started working at an orphanage. There was a lot of ground there that wasn't being put to use, so I started a small farm on the property. I first studied organic agriculture there, by reading books on my own. It was important to me that the kids grow up in a safe, natural environment."

Katsumi stayed at the orphanage for almost thirteen years, and then he and Ryoko decided to go to Irian Jaya in western New Guinea.

"We went as missionaries from a Japanese church," Katsumi explains. "West New Guinea, which was under Indonesian rule, was where we went to teach about organic agriculture to the native people. We stayed for just fourteen months."

Eventually, through their missionary group, they went to Campbell Farm in Wapato and fell in love with the area. After briefly staying at the farm, Katsumi worked as a gardener for three years, doing lawn care and maintenance. Then he and Ryoko purchased Mair Farm-Taki by getting financial help from relatives in Japan.

Ryoko, who has returned with a fresh pot of tea, tells me that she loves living in this area. "The diversity is wonderful, especially compared to Japan where everybody has black hair and the same skin color. Adding to the fact that Japanese people like to have control, they have lots of uniforms and rules. I never liked that. In America there is so much diversity. I like different races, a good mixture."

Katsumi agrees and then says almost proudly, "At the high school here it is almost 60 percent Hispanic, 16 percent Caucasian, 16 percent Native American, and the other 8 percent is Asian and Black."

The breadth of diversity in the area surprises me, and I had no idea there were any other Japanese.

Katsumi says, "Yes, but we didn't have any idea about that before we moved here. There are some families, but not as many as there used to be. Before World War II there were lots of Japanese in this area, and they were farming. Then they were sent to the camps and they lost all of their land. Only a few came back."

I know what Katsumi says is all too true. During my research about farms in Washington, I have learned that before

Fields That Dream

the war there were hundreds of Japanese farming across the state and thousands more in Oregon and California. The Japanese were vital to agriculture on the West Coast.

The internment of Japanese Americans during the war destroyed the lives of hundreds of thousands and forever changed the face of American agriculture. When the Japanese Americans were released, many farmers who had not been able to own the land they worked found their farms in ruins or taken over by their white neighbors. The fields were overgrown and the soil unturned, and a generation of work had been destroyed. Only a few Japanese Americans resurrected their farms, and now those generations of people are dying out.

Katsumi says mournfully, "Now there are only three or four Japanese families farming around here, and the children are going to town for jobs. The young generation doesn't like farming, they know it is very hard."

Ryoko agrees, "People want money, they want to have an easy life."

Katsumi continues, "When the first generation came from Japan, they were working, working, working. The second generation, they know that the first generation of farmers are very hardworking, and so they started to look for different jobs. Now, in the third generation, the farming is gone."

I ask if any one of their three children will want to continue the farm. They both laugh and shrug their shoulders. Katsumi says, "I don't know. I want them to, but I don't know."

Katsumi suggests that he take me on a tour of the farm, and we head outside. When I ask Ryoko if she will join us, she declines, smiling. "I have some work I need to finish up inside." She presses a jar of plum preserves into my hand as she hugs me good-bye.

Grafting

Katsumi and I put on our shoes and go outside into the crisp, gray air. He takes me around the pond and fields and introduces me to his goat and chickens. We end up in one of the orchards, where he shows me how he has grafted some apple trees.

I have heard of grafting before, and I know that it is a very common practice among farmers. But I am still amazed at the actual sight of it. Katsumi has taken a small branch from one kind of apple tree and literally stuck it to the trunk of another tree with a thick band of green wax. Eventually the new branch unites with the more established tree, and the tree starts producing the fruit of the new branch. It is a seamless transition. Katsumi smiles gently at my astonishment and explains that it really is an everyday procedure.

The sky has darkened slightly. I realize that it is getting late and I still have a long drive home. Katsumi escorts me to the car, and I take one last look at the farm before heading out.

On my way out of town it begins to dawn on me what extraordinary people the Takis are. Their kindness is gentle, and I can see how they are quietly upholding all that is dear to them. Mair Farm-Taki has created a memory of the Mairs, renewed the rich heritage of Japanese farming in the area, and nurtured a farm that needed love. But the Takis are also restoring their little piece of eastern Washington by breathing health and healing into a ground wracked with unanswered questions. It is as though they have been grafted to the land, and the soil swells to meet them, knowing it has been blessed.

Chapter Thirteen: The Heart of a Community

It is the end of June and I am headed to a church for a community meeting about the new light-rail system being planned for Seattle. The meeting concerns construction planning for the areas north of downtown, mainly around the University District and Northgate, the northernmost neighborhood within city limits. I have gulped down my dinner, afraid that my late entrance will be very noticeable in a rinky-dink meeting of neighborhood activists in the basement of a small church.

The church is, however, by no means small, and the huge parking lot is almost full. The Calvary Temple

reminds me, in this seemingly nonreligious country, that churches are still massive pillars in American neighborhoods and communities.

I have to take an elevator down to the basement, where there is a full-size basketball court that doubles as a community room. My late entrance is absolutely unnoticed in a crowd of more than 300 people. I am stunned; dinnertime on a weeknight, a meeting about boring old mass transit, and it is standing room only!

A week before I had not even known about the meeting, but last Saturday, on my weekly pilgrimage to the farmers' market, a huge yellow sign that said, "SAVE OUR MARKET!" accosted my eyes, and volunteers were urging shoppers to attend tonight's meeting.

The problem is as follows. After Seattle residents voted to plan for citywide light-rail, planners got to work and came up with five potential routes for north Seattle. However, four out of the five plans use the blacktop at the University Heights Center, where the farmers' market is held, as a staging area during construction. That means all of the dirt dug up during tunnel construction will be dumped there, and the site of the farmers' market becomes a big mud pit for at least two years. The light-rail route will also cut through a number of residential neighborhoods and severely impact the United Calvary Temple. Additionally, the farmers' market site doubles as a playground for a child-care center and a community garden, which would both have to be sacrificed as well. There is a fifth option, which bypasses the market site, neighborhoods, and church and obviously has the most community and city council support, but it is being seriously questioned by the Sound Transit Board because of cost.

The turnout at this Tuesday night meeting is impressive and heartening. There are a lot of residents of the potentially impacted area, local business owners, United Calvary Temple church members, and supporters of the University Heights Center. There are also city planners, architects, real estate mitigators, and environmental specialists on hand to facilitate the evening.

Each of our discussion groups agrees on a leader to present our opinions to the audience of community members and

Fields That Dream

planners. There is absolute consensus from the groups about the preferred route, for a number of reasons. All of the groups mention the importance of the farmers' market to the community.

Chris Curtis, the founder and manager of the University District Farmers' Market, presents the opinion of her discussion group to the audience. She is soft spoken in front of a large crowd, but direct and forceful in her presentation.

While talking about how the city has offered to move the market in compensation for the use of the site, she says, "The city doesn't understand what it takes to move a market, and they don't understand about the fragility of the University Business District."

The funky, somewhat rundown businesses on The Ave obviously greatly benefit from thousands of shoppers pouring into the area every Saturday over the course of the season. Chris also emphasizes the difficulty of effectively letting shoppers know about the new spot and then having to do it all over again when construction is over. Seeing Chris at the meeting, I realize the depth of her civic commitment and her determination to revitalize a struggling urban center. She is not only the manager of the market, she has become a vibrant community activist. This farmers' market is really not just for political diehards; it has become the heart of a very diverse neighborhood.

Farmers' markets across the country have had a similar impact. A large Saturday market in downtown Salt Lake City brings thousands of people to the financial district, which though bustling during the week is virtually deserted on the weekend. Moreover, the San Francisco farmers' market (one of the largest in the country with more than 150 vendors) is bringing new life to the recently renovated historic ferry terminal building. The city has decided to devote the building to

The Heart of a Community

businesses selling gourmet food and wine and kitchen supply stores such as Williams-Sonoma to complement what the market has to offer.

■ ◪ ▥ ▨

Chris didn't start the market with the intention of becoming a community activist; she was after a bit of simplicity. When I visit Chris at her home on the north shore of Lake Union to learn why she started the University District Farmers' Market more than a decade ago, she says wryly, "I guess it was an attempt to scale back my responsibilities and simplify my life after owning an ice cream shop on The Ave for many years."

Speaking of herself and her husband, Chris says, "We were anxious to get out of that business and change the direction of our lives." She laughs, "We must have been in some midlife crisis!"

Her initial plan to simplify her life may have backfired, seeing that Chris now manages the most financially successful farmers' market in the state and is on her way to creating a number of similarly successful markets in neighborhoods throughout the city. Starting the farmers' market, a nonprofit organization, was a true labor of love.

"I always knew I wanted to give back to the community in some way, I just put that on hold while I scrambled after money." She laughs, "After all, both my husband and I were part of the Vista program, a domestic Peace Corps, many years ago!"

She originally looked at the market as a way to renew an urban community, but after researching other farmers' markets, especially in California, she says, "That is when I started to understand the importance to farmers. Even though my motivation was community, I was learning more about the community of farmers and how they were this precious and rare commodity. Providing a good direct sales opportunity for them was a good thing to do."

Chris explains that direct sales at farmers' markets have become an essential part of the family farm income, and that, in a recent survey, 19,000 farmers reported selling their produce only at farmers' markets.[1]

Chris breaks down the economics even further. Shoppers spent $584.6 billion for food produced in the United States in 1998, with farmers earning only a 20 percent share of the food dollar. That is, farmers earned just 20 cents for every dollar spent on food. The other 80 percent went to food processing, packaging, shipping, and marketing costs.[2]

"At a farmers' market," Chris says with satisfaction, "100 percent of the profit goes to the farmer, minus the small cost of the annual market fee."

Despite the excellent opportunity for the farmers, Chris had to scour the area for funding and interested merchants. She asked the city for money, support, and the actual location for the market. "By hook and by crook we got that market open that first year for about $7,000. But we had to beat the bushes for farmers. I was helped incredibly by other markets and other market managers. I got close to 300 names of farmers. We sent out a survey, and we got eighty-five back. Of those eighty-five, forty said that they would sell at the market."

Chris had a meeting with the farmers who said they were interested in selling at the market but were wary that the focus would be more on crafts and premade food rather than agriculture. She nods appreciatively, recalling the gathering. "They were right. I don't know how many people have come up to me over the years and said, 'Thank you! This is about food, this is about farmers! Who needs more knickknacks anyway, you know?'"

The Heart of a Community

By the third week of June in 1993, she was ready to open with eighteen vendors, fourteen of whom were farmers. (A number of the original farmers are still there.) Chris says with a mix of incredulity and pride, "We didn't lose any farmers the first year, and by the end of the season we even had a few more. And the second season we had even more. And by the end of the second season we had spread out to the other half of the property. And by the fourth season we had a waiting list. The fifth and sixth season we had 40 percent growth."

The farmers' market is a relatively homegrown organization. Chris has two very part-time employees who help out on market days and for a few hours during the week. A lot of the work is done by a loyal group of volunteers, including her husband and daughter who get up at the crack of dawn on Saturday mornings to do any number of tasks such as setting up, pruning bushes, sweeping, and getting rid of mud puddles.

Trying to get a handle on why she has devoted her life to farmers' markets, I ask if Chris ever wanted to be a farmer herself.

"No way!" she says vehemently, shaking her head and chuckling. She explains that while she has farming in her blood, she never thought about making it her career.

Chris is from the Skagit Valley, about an hour north of Seattle, and her dad worked on a stud farm with pedigree bulls.

"I grew up with kids who were the children of farmers, and many of them took on that responsibility. They went to college in agriculture and then they would come back and take on the family farms. And that is what they still do. You go to high school reunions and almost everyone is farming.

"But I wanted to leave. I wanted to grow up and go to college and go away to a city and stuff, which is pretty much what I did."

Fields That Dream

Chris's husband, Tim, strides into the room and sits down on the couch next to me. He is tall, lanky, and covered with dirt after having spent the whole afternoon in the garden. While I have Tim in the room, I ask him why he likes helping out at the farmers' market.

He answers matter-of-factly, "It is infectious. It is not like lugging boxes at a warehouse or something. We joke that it is our weekly cocktail party without the cocktails. I mean, you go there, you talk to all these people who all share an interest in good food."

Tim leans forward and says thoughtfully, "It really has, on many levels, returned to the retailing formula of a hundred years ago. The person who grows it sells it ... people take their own bags and meet friends there. No plastic carts. Sometimes when I see farmers with a cell phone I laugh to myself, 'You are not supposed to have that! It is bad for the image!' We noticed in the first year that the farmers would go to great lengths to have typed up their price cards and laminate them. But people don't want that. They want prices written on the back of a paper bag."

Chris relates to me how large supermarkets are aiming to capitalize on that very same farmers' market experience, which harkens back to a more congenial and simple time. "Larry's Markets is now using wooden bins to hold their produce and posting signs about where the farms are and who grows the food. And," Chris says with incredulity, "a large grocery store in West Seattle actually wants to have a small farmers' market in their store parking lot. The store representative who got in touch with me said, 'Yeah, we want to do that because, you know, it's all about the farmer.'"

"The farmer?" Her voice goes up an octave. "I went, wait a minute ... the farmer. You represent a huge grocery chain. You

The Heart of a Community

are more about something else. But, I suppose I am glad he felt that way ... I mean, whatever the motivation, this is all good, connecting consumers to their producers."

Time and time again during our conversation, Chris comes back to how important the market is on many different levels. "I find inspiration in the farmers themselves, visiting farms, discovering what it is they do, and the importance of what they do. I don't know that I had given all this much thought when I first thought about starting the market. I have to say, the farmers have put me on a steep learning curve about all of this. A lot of the work that I do is outside of the details of the markets. I work on a lot of farm issues. I sit on a ton of boards, like the King County Agricultural Commission and the State Association of Farmers' Markets.

"I feel real good that there are some huge public benefits to all this. That was the direction I was headed, and it is working on so many levels, on a community level, farming level, food level, and saving-agricultural-land level. It is good to know that the work you do is of benefit."

Chris is quick to point out that many of the great things that come out of the market have nothing to do with her. She stresses the importance of the farmers, shoppers, and volunteers. She says to me, almost urgently, "There is something I really want you to know about. Without fail every Saturday afternoon, two volunteers from the food bank show up at the market and they go around with their hand truck and bin and collect a lot of produce from the farmers. They take it straight away back to the food bank to process it. Over 10,000 pounds of fresh produce went to the food bank last summer. It is a wonderful link."

The University District Farmers' Market is by no means

Fields That Dream

alone in its food recovery efforts, another sign that farmers' markets are becoming integral parts of the communities they serve. While the USDA estimates that more than one quarter of all the food produced in the United States is thrown away, farmers' markets are actively *not* part of that problem. Market coordinators across the country are collaborating with food banks and charitable organizations to donate unsold produce at the end of the selling day. A report written by analysts Charlene Price and J. Michael Harris explains that the "USDA initiated a pilot program in 1997 which matched producers in the farmers' market setting with nonprofit food recovery and gleaning organizations in the Washington, DC, area. Over 8,000 pounds of food were recovered during 3 months of 1997 and over 12,000 pounds of food during 5 months of 1998."[3] Furthermore, 58 percent of markets participate in WIC (Women with Infants and Children) coupons, food stamps, and local and/or state nutrition programs.[4]

With these kinds of partnerships in place, Chris feels that the future of farmers' markets in Seattle is bright. Her idea is to use the model of the University District Market in neighborhoods throughout the city. So far she has started a new market in West Seattle and is collaborating with a market manager in a south Seattle neighborhood. "Eventually, I would like to see a consolidation of markets, a centralized organization, and wages for the employees and managers of the market commensurate with work. There are a lot of neighborhoods clamoring for them. And the response has been wonderful in Columbia City and West Seattle.

"I think it would be entirely possible to double the number of farmers' markets in the city without doubling the work, and I definitely think I will stick with it.

The Heart of a Community

"I love the market. I mean, I do love the actual event itself. It is very satisfying. It comes together, it looks beautiful ... smells wonderful ... "

Chris pauses, wiggles her shoulders a little, as though the weight of the matter is settling down on her, "and, it's funny, I really and truly was thinking of the farmers' market as a catalyst for the community, and I wasn't thinking about the community of farmers either. But then it didn't take long for me to figure out how important this is to people."

She is comfortable with this realization and the responsibility she has to the farmers and their urban neighbors. I think Chris will do about damn near anything to keep this market going, even if the farmers have to sell their produce on the back of a bulldozer that is dumping dirt onto the University Heights blacktop.

Epilogue

The University District Farmers' Market was central to my life in Seattle and central in my political, intellectual, and creative development as a writer, oral historian, and activist. This book is like a photo album of market memories. Rich colors, sunlight, plate-sized dahlias, and mountains of ruby red tomatoes. Laughter. Faces drenched in summer and long, work-filled days. Kitchen tables, greenhouses, and conversations caught on paper the way breath, in the dead of winter, freezes in midair.

I think about the farmers who opened their homes, took time out of their days to tromp through the fields with me, my tape recorder in hand. I think about the effort they made to go over the words I had written about them, making sure I got the details right. And I think about how wonderful it was knowing the growers by name, asking about their kids while deliberating whether to get butterhead or oak leaf lettuce.

After six years of living in Seattle, I have moved "home" to California. My partner and I live in Santa Cruz, forty-five minutes north of Pacific Grove, my hometown. At the new market I go to in Santa Cruz, I'm beginning to get to know the farmers, and they remind me of the growers in Seattle: dedicated, hardworking, and deeply principled about being stewards of the land. I have no doubt that each grower in every market around the country has an extraordinary story, because, in many ways, the University District Farmers' Market in Seattle, Washington, is like any other farmers' market in the United States. There *are* market managers like Chris Curtis all over the country who are striving to make a connection between growers and consumers and working to bridge the ever-widening gap between rural

and urban communities. There *are* farmers like Joanie McIntyre and Michael Shriver, raising their children with the small-
farm values of hard work and respect for the land. And, there
are farmers like John Huschle, who has turned his own farm
into a personal revolution against globalization.

We, the consumers, are essential in supporting these farmers
and markets. We can truly be part of the movement that supports sustainable and locally grown food. In a world where we
often feel helpless and overwhelmed, remember that ultimately,
we hold the power, for in a market economy it is the consumers
who have the final say. As the Chef's Collaborative from Boston
proclaims, "Vote with your fork for a sustainable future."[1]

Our voices are being heard. Since 1999, tens of thousands
of people have continued to gather wherever the World Trade
Organization has met to protest the policies of world trade.
Ninety-four percent of Americans want labels on genetically
engineered foods[2] and the European Union worked hard to ban
GMOs entirely. Even fast-food companies, under pressure from
consumer demand, are offering healthy options such as salads
and low-fat sandwiches.

Things are even beginning to shift within the organics
movement. Because there is no guaranteed connection between
organic standards and labor standards, four international environmental and human rights groups have teamed up to create
the Social Accountability in Sustainable Agriculture (SASA)
program.[3] "The intent of the collaborative effort is to set international standards for the treatment of farmworkers and to
develop a certification to guarantee retailers and consumers
that the farm products they buy were grown and picked by
well-treated workers."[4]

By purchasing our food at a local farmers' market, we are

Fields That Dream

adding our voices to this global chorus for change. Shopping becomes a political act and eating locally grown organic food provides a practical way of saying "No." No to the massive movement to globalize the world economy. No to Monsanto and genetically modified food. No to poisoning the ecosystem. And no to strip malls, gulping up farmland with an insatiable hunger.

We are also saying yes. When we buy locally and sustainably grown food we are honoring the farmers, we are honoring our future, and we are honoring the earth.

I take my one-year-old son to the farmers' market now. He toddles around, looking wide-eyed at all of the other children, the colors, and the bustle of activity. He stops to bounce to the music of the slide guitarist and gnaws on the free apple sample. I feel so blessed to have this beautiful healthy food for him to eat and so thankful to the farmers for growing it. May we all grow and thrive like my son, who eats the bright orange carrots of California. May all our hopes for this planet of ours, this beautiful home, take root in healthy soil and sprout for generations and generations and generations to come.

Resources

Organizations:

Farmland Preservation

American Farmland Trust
http://www.farmland.org
1200 18th St. NW
Washington, DC 20036
(202) 331-7300
Fax: (202) 659-8339
info@farmland.org

PCC Farmland Fund
http://www.pccnatural
markets.com/farmlandfund
4201 Roosevelt Way NE
Seattle, WA 98105
(206) 547-1222, ext. 140
Fax: (206) 545-7131
farmlandfund@pccsea.com

Pesticide Awareness

National Pesticide
Information Center
http://npic.orst.edu
(800) 858-7378

Pesticide Action Network
http://www.panna.org
49 Powell St., Ste. 500
San Francisco, CA 94102
(415) 981-1771
Fax: (415) 981-1991
panna@panna.org

Farmers' Markets Directories

Local Harvest
http://www.localharvest.org
Ocean Groups Inc.
Santa Cruz, CA
(831) 475-8150

United States
Department of Agriculture
http://www.ams.usda.gov/
farmersmarkets

Agricultural Workers' Rights

Global Exchange
http://www.globalexchange.org
2017 Mission St., #303
San Francisco, CA 94110
(415) 255-7296

Social Accountability in
Sustainable Agriculture
http://www.isealalliance.
org/sasa

United Farm Workers
http://www.ufw.org
National Headquarters –
UFW
P.O. Box 62
Keene, CA 93531

Education

Junior Master Gardener
http://jmgkids.com
225 Horticulture/
Forestry Bldg.
Texas A&M University
College Station, TX
77843-2134
(979) 845-8565
Fax: (979) 845-8906
programinfo@jmgkids.org

National Gardening
Association
http://www.garden.org/home
http://www.kidsgardening.com
1100 Dorset St.
South Burlington, VT 05403
(802) 863-5251

Smithsonian Institution –
Seeds of Change Garden
http://www.mnh.si.edu/
archives/garden

Humane Farming

Food Animal Concerns Trust
http://www.fact.cc
P.O. Box 14599
Chicago, IL 60614
(773) 525-4952
Info@FACT.cc

Humane Farming Association
http://www.hfa.org
P.O. Box 3577
San Rafael, CA 94912
(415) 771-CALF
Fax: (415) 485-0106
hfa@hfa.org

Sustainable Agriculture

American Community
Gardening Association
http://www.organic
consumers.org
http://www.community
garden.org
c/o Council on the Environment of New York City
51 Chambers St., Ste. 228
New York, NY 10007
(877) ASK-ACGA or
(212) 275-2242

Community Alliance with
Family Farmers
http://www.caff.org
P.O. Box 363
Davis, CA 95617
(530) 756-8518
Fax: (530) 756-7857

Ecological Farming
Association
http://www.eco-
farm.org/index.html
406 Main St., Ste. 313
Watsonville, CA 95076
(831) 763-2111
Fax: (831) 763-2112
info@eco-farm.org

Organic Consumers
Association
6101 Cliff Estate Rd.
Little Marais, MN 55614
(218) 226-4164
Fax: (218) 353-7652

Organic Farming Research
Foundation
http://www.ofrf.org
P.O. Box 440
Santa Cruz, CA 95060
(831) 426-6606
Fax: (831) 426-6670
info@ofrf.org

Genetically Modified Food

Council for Responsible
Genetics
http://www.gene-watch.org
5 Upland Rd., Ste. 3
Cambridge, MA 02140
(617) 868-0870

Union of Concerned
Scientists
http://www.ucsusa.org
2 Brattle Sq.
Cambridge, MA 02238-9105
(617) 547-5552
Fax: (617) 864-9405

Seed Saving

International Seed Saving
Institute
http://www.seedsave.org
P.O. Box 4619
Ketchum, ID 83340
(208) 788-4363
Fax: (208) 788-3452

Seed Savers Exchange
http://www.seedsavers.org/
Home.asp
3094 N. Winn Rd;.
Decorah, IA 52101
(563) 382-5990
Fax: (563) 382-5872

Recommended Reading:

Ableman, Michael. *From the Good Earth: A Celebration of Growing Food Around the World.* New York: Harry N. Abrams, Inc., 1993.

———. *On Good Land: The Autobiography of an Urban Farm.* San Francisco: Chronicle Books, 1998.

Fowler, Cary, and Pat Mooney. *Shattering: Food, Politics, and the Loss of Genetic Diversity.* Tucson: The University of Arizona Press, 1990.

Hein, Teri. *Atomic Farmgirl: Growing Up Right in the Wrong Place.* New York: Mariner Books, 2003.

Henderson, Elizabeth, and Robyn Van En. *Sharing the Harvest: A Guide to Community Supported Agriculture.* White River Junction, Vt.: Chelsea Green Publishing Company, 1999.

Jackson, Wes. *Becoming Native to This Place.* New York: Counterpoint Press, 1996.

Lewis, Tom. *Divided Highways: Building the Interstate Highways, Transforming American Life.* New York: Viking Penguin, 1997.

Nabhan, Gary Paul. *Coming Home to Eat: The Pleasures and Politics of Local Foods.* New York: W.W. Norton & Company, 2002.

Nestle, Marion. *Food Politics: How the Food Industry Influences Nutrition and Health.* Berkeley: University of California Press, 2002.

Robbins, John. *The Food Revolution: How Your Diet Can Help Save Your Life and Our World.* York Beach, Maine: Conari Press, 2001.

Rothenberg, Daniel. *With These Hands: The Hidden World of Migrant Farmworkers Today.* New York: Harcourt Brace and Company, 1998.

Schlosser, Eric. *Fast Food Nation: The Dark Side of the All-American Meal.* New York: Perennial, 2002.

Thomas, Janet. *The Battle in Seattle: The Story Behind and Beyond the WTO Demonstrations.* Golden, Colo.: Fulcrum Publishing, 2000.

Wilson, Duff. *Fateful Harvest: The True Story of a Small Town, a Global Industry, and a Toxic Secret.* New York: Harper Collins Publishers, 2001.

Wirzba, Norman. *The Essential Agrarian Reader: The Future of Culture, Community, and the Land.* Lexington: University of Kentucky Press, 2003.

Endnotes

Introduction: Beginnings

[1] Northwest Coalition for Alternatives to Pesticides (NCAP), "What Is a Pesticide?" *Journal of Pesticide Reform* 2, no. 19 (summer 1999): 2.

[2] Michael Ableman, *From the Good Earth: A Celebration of Growing Food Around the World* (New York: Harry N. Abrams, Inc., 1993), 74.

[3] Pesticide Reduction Initiative, "The Need for Change," http://www.smallparty.org/reducepesticides/learnmore/need_ for_change.html.

[4] Ibid.

[5] Ibid.

[6] American Farmland Trust, Farmland Information Center, "Fact Sheet: Why Save Farmland?" http://www.farmlandinfo.org/ documents/28562/FS_Why%20Save%20Farmland_1-03.pdf.

[7] Ibid.

[8] Ibid.

Chapter 1: Pioneer Roots

[1] Richard Kirkendall, "Up to Now: A History of American Agriculture from Jefferson to Revolution to Crisis," *Agriculture and Human Values* 4, no. 1 (winter 1987): 5.

[2] Richard White, *"It's Your Misfortune and None of My Own": A New History of the American West* (Norman: University of Oklahoma Press, 1991), 137.

[3] Ibid., 143.

[4] Kate Miller, "Brief Summary of Native American History: 1830–1910" (lecture, Monterey Peninsula College, Monterey,

Calif., 20 November 2003).

5 Elizabeth Cook-Lynn, *Why I Can't Read Wallace Stegner and Other Essays: A Tribal Voice* (Madison: the University of Wisconsin Press, 1996), 29.

Chapter 2: Good Girl/Earth Mama

1 Michael Ableman, *From the Good Earth: A Celebration of Growing Food Around the World* (New York: Harry N. Abrams, Inc., 1993), 70.

2 Craig Canine, *Dream Reaper: The Story of an Old-Fashioned Inventor in the High Tech, High Stakes World of Modern Agriculture* (New York: Alfred A. Knopf, 1995), 191.

3 David D. Danbom, *Born in the Country: A History of Rural America* (Baltimore: Johns Hopkins University Press, 1995), 236–237.

4 Richard Kirkendall, "Up to Now: A History of American Agriculture from Jefferson to Revolution to Crisis," *Agriculture and Human Values* 4, no. 1 (winter 1987): 10.

5 Joan Dye Gussow, *Chicken Little, Tomato Sauce and Agriculture: Who Will Produce Tomorrow's Food?* (New York: The Bootstrap Press, 1991), 4.

6 Gary Paul Nabhan, *Coming Home to Eat: The Pleasures and Politics of Local Foods* (New York: W.W. Norton & Company, 2002), 61.

7 Desmond Jolly, "The Current State of U.S. Agriculture" (lecture, Ecological Farming Conference, Asilomar Conference Center, Pacific Grove, Calif., 24 January 2002).

8 Gilbert C. Fite, *American Farmers: The New Minority* (Bloomington: Indiana University Press, 1981), 102–119.

9 Gene Logsdon, "All Flesh Is Grass: A Hopeful Look at the Future of Agrarianism" in *The Essential Agrarian Reader: The*

Fields That Dream

Future of Culture, Community, and the Land, ed. Norman
Wirzba, 154–170 (Lexington: University of Kentucky Press,
2003), 159.

[10] Associated Press, "Corporate Ag Tries to Justify Hogging
Government Subsidies," *Las Vegas SUN*, 10 September 2001,
(accessed through the Organic Consumers Association,
http://www.organicconsumers.org/corp/subsidies091201.cfm).

[11] United States Department of Agriculture, "The National
Organic Program – Background and History,"
http://www.ams.usda.gov/nop/Consumers/background.html.

[12] Mara Dyczewski, "How to Comment on the Proposed National
Organic Standards," *Seattle Tilth* 22, no. 5 (June 2000): 1.

Chapter 3: Natural Progression

[1] Tom Lewis, *Divided Highways: Building the Interstate Highways,
Transforming American Life* (New York: Viking Penguin, 1997),
90–91.

[2] Eric Schlosser, *Fast Food Nation: The Dark Side of the All-
American Meal* (New York: Perennial, 2002), 22.

[3] Lewis, *Divided Highways*, 286.

[4] Ibid., x.

[5] David Halberstam, *The Fifties* (New York: Villard Books, 1993),
134–137.

[6] Lewis, *Divided Highways*, 71.

[7] Halberstam, *The Fifties*, 157.

[8] Schlosser, *Fast Food Nation*, 4.

Chapter 4: Moving Easy in Harness

[1] Erik Marcus, *Vegan: The New Ethics of Eating* (Ithaca, N.Y.:
McBooks Press, 1998), 125.

[2] Ibid., 96.

[3] R. W. Trullinger, "Science in the Agriculture of Tomorrow," in *Crops in Peace and War: Yearbook of Agriculture 1950–1951* (Washington, D.C.: United States Government Printing Office, 1951), 1.

[4] John Robbins, *The Food Revolution: How Your Diet Can Help Save Your Life and Our World* (York Beach, Maine: Conari Press, 2001), 234.

[5] Ibid., 236.

[6] Ibid., 241.

[7] Ibid., 256.

[8] Lisa M. Hamilton, "From Nature to Science to Sickness to Health: Coming Back to Organic Meat," *The Newsletter of CCOF — California Certified Organic Farmers* 19, no. 4 (winter 2002–2003): 3.

[9] Ibid.

Chapter 5: Backyard Homestead

[1] Janet Thomas, *The Battle in Seattle: The Story Behind and Beyond the WTO Demonstrations* (Golden, Colo.: Fulcrum Publishing, 2000), 49.

[2] Ibid., 50.

[3] Kevin Danaher and Jason Mark, *Insurrection: Citizen Challenges to Corporate Power* (New York: Routledge, 2003), 241.

[4] Ibid., 242–243.

[5] World Trade Organization, "The World Trade Organization …" http://www.wto.org/english/res_e/doload_e/inbr_e.pdf, 7.

[6] Ibid., 4–5.

[7] Danaher and Mark, *Insurrection*, 248.

[8] Northern Plains Resource Council, "The World Betrayed by the World Trade Organization," *Northern Plains Resource Council Newsletter*, (November 1999): 1.

[9] Peter Rosset, "A New Food Movement Comes of Age in Seattle," in *Globalize This!: The Battle Against the World Trade Organization and Corporate Rule*, ed. Kevin Danaher and Roger Burbach (Monroe, Maine: Common Courage Press, 2000), 137.

[10] Vandana Shiva, "The Historic Significance of Seattle," *Alliance for Sustainable Jobs and the Environment* 2, no. 1 (winter 1999): 16–17.

[11] Frederick Merk, *Manifest Destiny and Mission in American History: A Reinterpretation* (New York: Vintage Books, 1963), 24.

[12] TransFair USA, "What Is Fair Trade Certification?" http://www.transfairusa.org/content/about/certification.php.

Chapter 6: Borders

[1] Gloria Anzaldùa, *Borderlands, La Frontera: The New Mestiza* (San Francisco: Aunt Lute Books, 1987), 25.

[2] Margaret Reeves, Anne Katten, Martha Guzmán, "Fields of Poison 2002: California Farmworkers and Pesticides," United Farmworkers of America, http://www.ufw.org/reportexsum.pdf, 4.

[3] Ibid.

[4] Daniel Rothenberg, *With These Hands: The Hidden World of Migrant Farmworkers Today* (New York: Harcourt Brace and Company, 1998), 19.

[5] Stephen H. Sosnick, *Hired Hands: Seasonal Farm Workers in the United States* (Santa Barbara, Calif.: McNally & Loftin, West, 1978), 54–55.

[6] Rothenberg, *With These Hands*, 36.

[7] Ibid., 6.

[8] Hondagneu-Sotelo Pierette, "Unpacking 187: Targeting

Mejicanas," in *Immigration and Ethnic Communities: A Focus on Latinos*, ed. Refugio I. Rochin, (East Lansing, Mich.: Julian Samora Research Institute, 1996), 93–101.

[9] Ibid., 94.

[10] Rothenberg, *With These Hands*, 135.

[11] Ibid.

[12] Ibid., 126.

[13] Eduaro Porter, "Illegal Immigrants Are Bolstering Social Security with Billions," *The New York Times*, 5 April 2005, sec. A1.

[14] Ibid.

[15] Lynda V. Mapes, "Illegal, but Needed, Workers Gaining Ground," *Seattle Times*, 18 June 2000, sec. A1.

Chapter 7: Mountain People

[1] Nancy D. Donnelly, *Changing Lives of Refugee Hmong Women* (Seattle: University of Washington Press, 1994), 4.

[2] Phuong Le, "Hmong Carving out a New Landscape," *Seattle Post Intelligencer*, 5 October 2000, sec. A1.

[3] Lillian Faderman and Ghia Xiong, *I Begin My Life All Over: The Hmong and the American Immigrant Experience* (Boston: Beacon Press, 1998), 3.

[4] Ibid., 6.

[5] Ibid., 7.

[6] Lao Human Rights Council, "Appeal to the U.S. Government to Stop the Ethnic Cleansing War and Biological and Chemical Warfare in Laos," http://www.laohumrights.org/terror.html.

[7] Cary Fowler and Pat Mooney, *Shattering: Food, Politics, and the Loss of Genetic Diversity* (Tucson: The University of Arizona Press, 1990), 56.

[8] Ibid., 59.

Fields That Dream

[9] Kristina Canizares, "Hmong American Community," Oxfam America, http://www.oxfamamerica.org/global/art4163.html.

[10] Vang Pobzeb, "2001 Hmong Population and Education in the United States and the World," Lao Human Rights Council, Inc., http://www.laohumrights.org/2001data.html.

Chapter 8: Snow Geese

[1] Cary Fowler and Pat Mooney, *Shattering: Food, Politics, and the Loss of Genetic Diversity* (Tucson: The University of Arizona Press, 1990), 26.

[2] Dan Whipple, "Analysis: Treaty Favors GM Crop Protection," *Washington Times*, 5 April 2004. http://www.washtimes.com/upi-breaking/20040402-115247-6665r.htm.

[3] Kim Waddell, e-mail message to author, 28 August 2004.

[4] Fowler and Mooney, *Shattering*, 45.

[5] Whipple, "Analysis," 2.

[6] Puget Consumers Co-op. "The Reality of GMOs" pamphlet (May 2000), 3.

[7] John Robbins, *The Food Revolution: How Your Diet Can Help Save Your Life and Our World* (York Beach, Maine: Conari Press, 2001), 315.

[8] Paul Raeburn, "After Taco Bell: Can Biotech Learn Its Lesson?" *Business Week*, 6 November 2000, 54.

[9] Kim Waddell, e-mail message to author, 28 August 2004.

[10] Robbins, *The Food Revolution*, 309.

[11] Ibid., 311.

[12] Kim Waddell, e-mail message to author, August 28, 2004.

[13] Union of Concerned Scientists, "Questions and Answers on *Gone to Seed*," http://www.ucsusa.org/food_and_environment/biotechnology.

[14] Margaret Mellon and Jane Rissler, *Gone to Seed: Transgenic*

Contaminants in the Traditional Seed Supply (Union of Concerned Scientists, 2004), 1.

[15] Fowler and Mooney, *Shattering*, 63–65.

[16] Seed Savers Exchange, "What Are Heirlooms," http://www.seed savers.org/savingheirlooms.html.

Chapter 9: Second Chances

[1] National Coalition for the Homeless, "Homeless Youth: NCH Fact Sheet #11," http://www.nationalhomeless. org/youth.html, 1.

[2] Michael Ableman, *From the Good Earth: A Celebration of Growing Food Around the World* (New York: Harry N. Abrams, Inc., 1993), 122.

[3] Joan White (National Gardening Association), in discussion with the author, January 2004.

[4] Aarti Subramaniam, "Garden-Based Learning in Basic Education: A Historical Review," *Monograph* (summer 2002), http://fourhcyd.ucdavis.edu/extending/pubs/focus/pdf/MO02 V8N1.pdf.

[5] Ibid., 6.

Chapter 10: The Back Forty ... Inches

[1] Allan D. Wallis, *Wheel Estate: The Rise and Decline of Mobile Homes* (New York: Oxford University Press, 1991), 64.

[2] Marion Nestle, *Food Politics: How the Food Industry Influences Nutrition and Health* (Berkeley: University of California Press, 2002), 179.

[3] Ceci Connolly, "Public Policy Targeting Obesity," *Washington Post*, 10 August 2003, sec. A1.

[4] Eric Schlosser, *Fast Food Nation: The Dark Side of the All-American Meal* (New York: Perennial, 2002), 47–49.

Fields That Dream

[5] Nestle, *Food Politics*, 7.

[6] Michael Ableman, *On Good Land: The Autobiography of an Urban Farm* (San Francisco: Chronicle Books, 1998), 112.

Chapter 11: Sky's the Limit

[1] Michael Pollan, "Behind the Organic-Industrial Complex," *New York Times Magazine*, 13 May 2001 (accessed through http://www.mindfully.org/Food/Organic-Industrial-Complex.htm).

[2] Ibid.

[3] Marion Nestle, *Food Politics: How the Food Industry Influences Nutrition and Health* (Berkeley: University of California Press, 2002), 13.

[4] Michael Pollan, "Behind the Organic-Industrial Complex," *New York Times Magazine*, 13 May 2001 (accessed through http://www.mindfully.org/Food/Organic-Industrial-Complex.htm).

Chapter 12: Grafting

[1] Duff Wilson, *Fateful Harvest: The True Story of a Small Town, a Global Industry, and a Toxic Secret* (New York: Harper Collins Publishers, 2001), 11.

[2] Ibid.

[3] Ibid.

[4] Michael Ableman, *From the Good Earth: A Celebration of Growing Food Around the World* (New York: Harry N. Abrams, Inc., 1993), 86.

[5] Ibid.

[6] Teri Hein, *Atomic Farmgirl: Growing Up Right in the Wrong Place* (New York: Mariner Books, 2003), xi.

[7] Ibid., 249.

[8] Paul Koberstein, "The Sacrifice Zone," *Cascadia Times* (winter 2004), http://www.times.org/archives/2004/sacrificezone.htm.

[9] Terry Tempest Williams, *Refuge: An Unnatural History of Family and Place* (New York: Vintage Books, 1991), 284.

[10] Dr. Bob Stevens, telephone interview, 26 July 2004.

Chapter 13: The Heart of a Community

[1] United States Department of Agriculture, "Farmers Market Facts!" http://www.ams.usda.gov/farmersmarkets/facts.htm.

[2] Marion Nestle, *Food Politics: How the Food Industry Influences Nutrition and Health* (Berkeley: University of California Press, 2002), 18.

[3] C. Price and J. Michael Harris, "Increasing Food Recovery From Farmers' Markets: A Preliminary Analysis," Charlene Economic Research Service/United States Department of Agriculture, http://www.ers.usda.gov/publications/FANRR4/.

[4] United States Department of Agriculture, "Farmers Market Facts!"

Epilogue

[1] Marion Nestle, *Food Politics: How the Food Industry Influences Nutrition and Health* (Berkeley: University of California Press, 2002), 373.

[2] Al Krebs, "New Poll—94% of Americans Want Labels on GE Foods," *Agribusiness Examiner* 295 (19 October 2003), (accessed through Organic Consumers Association, http://www.organicconsumers.org/ge/newpoll102303.cfm.)

[3] Laurel Chesky, "Good Works: Local Strawberry Farm Serves as 'Guinea Pig' for Development of International Farm Labor Standards," *Santa Cruz* (Calif.) *Good Times*, 17–23 July 2003, 12.

[4] Ibid.

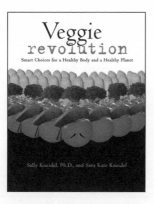
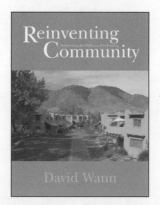